Adhesives, Coatings and Consolidants

This new edition titled *Adhesives, Coatings and Consolidants* discusses the chemical, physical, and optical properties that influence the selection of adhesives, coatings, and consolidants used in the conservation of cultural heritage objects made from a wide range of materials.

The structured text guides readers through the concepts that underpin the selection of the material to use for a given object, when making a join, applying a coating, or consolidating a fragile surface. Widely used conservation materials today are mentioned in the context of the desirable properties that account for their popularity, while examples of those used in the past and likely to be found on previously treated objects are also discussed. This volume is not a recipe book. It is a basic primer and an aid to the assimilation of concepts that will become embedded in decision-making by conservation professionals. Adhesives are discussed first and in the most detail, followed by coatings. The section on consolidants has been extended to reflect developments in treatment and ongoing research into more effective and more sustainable products, such as nanomaterials, that offer new possibilities for the treatment of challenging objects. A list of further reading and some practical exercises are included, along with a comprehensive index.

Thoroughly updated to reflect the most recent research, *Adhesives, Coatings and Consolidants* remains a crucial resource for those beginning a course of study in practical conservation and as a refresher for emerging and mid-career conservators.

Joyce H. Townsend, PhD, FIIC, ACR, is Series Editor for the updated *Science for Conservators* series. She has been IIC Director of Publications for over ten years and a conservation scientist for over four decades, more than three of them spent at Tate in London, UK, where she specialises in microscopical methods for the analysis of paint and the interpretation of the techniques of British artists working in oil and watercolour, as well as microfading and X-ray fluorescence studies for works on paper. She has published widely for a range of audiences. Since 2019 she has been an honorary professor in the School of Culture and Creative Arts, University of Glasgow, UK.

Science for Conservators
Series editor: Joyce H. Townsend

The *Science for Conservators* series is now being revised and fully updated, and it will be extended in scope. The original series has provided key basic texts for conservators throughout the world since its publication in the late 1980s. Scientific concepts are basic to the conservation of artefacts of every type, and these introductory volumes provide an essential theoretical background for conservators who have entered the field without scientific training. It is designed for pre-programme, student and personal study, and also serves as a checklist of scientific terms for those working in English as a second or third language.

An Introduction to Materials and Chemistry
Joyce H. Townsend

Colour, Colour Measurement and Colour Change
David Saunders

Experimental Design and Scientific Data Analysis
Lynda K. Skipper and Philip J. A. Skipper

Adhesives, Coatings and Consolidants
Joyce H. Townsend

For more information about this series, please visit: www.routledge.com/Science-for-Conservators/book-series/SFC

Adhesives, Coatings and Consolidants

Third Edition

Joyce H. Townsend

Routledge
Taylor & Francis Group
LONDON AND NEW YORK

Designed cover image: Plaster consolidation of a 20th-century mosaic on the Buzludzha Monument, Bulgaria, during emergency conservation treatment by the Buzludzha Project Foundation in 2020. Image: Nikifor Haralampiev. Reproduced with the permission of Nikifor Haralampiev.

Third edition published 2025
by Routledge
4 Park Square, Milton Park, Abingdon, Oxon, OX14 4RN

and by Routledge
605 Third Avenue, New York, NY 10158

Routledge is an imprint of the Taylor & Francis Group, an informa business

First edition published by the Crafts Council 1983
Second edition published by Routledge 1992

British Library Cataloguing-in-Publication Data
A catalogue record for this book is available from the British Library

Library of Congress Cataloging-in-Publication Data
Names: Townsend, Joyce H., editor.
Title: Adhesives, coatings and consolidants / Joyce H. Townsend.
Description: Third edition. | Abingdon, Oxon ; New York, NY : Routledge, 2025. | Series: Science for conservators | Includes bibliographical references and index.
Identifiers: LCCN 2024036831 (print) | LCCN 2024036832 (ebook) | ISBN 9781032484624 (hardback) | ISBN 9781032484617 (paperback) | ISBN 9781003389217 (ebook)
Subjects: LCSH: Adhesives. | Coatings. | Art—Conservation and restoration.
Classification: LCC TP968 .A288 2025 (print) | LCC TP968 (ebook) | DDC 702.8/8—dc23/eng/20241021
LC record available at https://lccn.loc.gov/2024036831
LC ebook record available at https://lccn.loc.gov/2024036832

ISBN: 978-1-032-48462-4 (hbk)
ISBN: 978-1-032-48461-7 (pbk)
ISBN: 978-1-003-38921-7 (ebk)

DOI: 10.4324/9781003389217

Typeset in Times New Roman
by Apex CoVantage, LLC

Contents

Figures

Tables

Foreword

Using adhesives, coatings, and consolidants is inherently interventive. The materials chosen for such treatments are likely to influence the long-term survival of the object and in some cases its appearance, both immediately and far into its future as well. The main aims of this book, in the *Science for Conservators* series, are to help you appreciate the underlying science involved in the joining, coating, and consolidation of objects and to provide tools for understanding the properties of the various types of material you might select or reject for a treatment. Chemistry underlies some of these processes, but physics and materials testing also play a significant part. A comprehensive evaluation of the performance of an adhesive, coating, or consolidant might even involve optical metrology (the measurement of things using visible light), or investigation of its potential to suffer from biodeterioration in buildings or outdoor sites, or investigations more typically used in other scientific fields such as mechanical engineering.

In contrast to cleaning (the subject of *Cleaning* in this series) which includes the *removal* of the material products of age and environment from an object, joining, coating, and consolidation depend on the *addition* of new material. Admittedly the use of an adhesive or coating may be preceded by removal of an older organic material which was no longer working efficiently at joining components or protecting a surface, which had aged beyond its useful lifetime, or which proved so unsuitable for the purpose (or the object's environment) that it endangered the survival of the object. Previous consolidants are not usually treated with the aim of removing them before another cycle of consolidation, for the good reason that their complete extraction is impossible in practice. Even partial removal is a challenge, and past consolidation tends to narrow present and future treatment options considerably, sometimes drastically, and more so than processes to re-attach fragments together.

The three processes of joining fragments, coating for protection, and consolidating surfaces to prevent ongoing loss of original material have other features in common, which makes it very useful to consider them together. First, the materials used as adhesives, coatings, or consolidants usually need to meet similar technical requirements. For instance, to work effectively, a coating or consolidant should behave like an adhesive in that it should stick to the surface of the object with which it is in contact; much of the scientific understanding of how materials stick together is common to all three processes. Second, many of the materials used are polymers (very large molecules) of one sort or another. By studying the structure and properties of polymers in general, you can learn a lot about the effectiveness of particular polymeric materials in each of the three processes.

In defining the scope of this book, the properties of the materials *added* to an object have been emphasised, but this is only part of the technical side of the problem. The addition of any material to an object will affect the object in some way, and this can raise a whole range of ethical problems. Can even an adhesive be removed without damaging the original object? It may, for example, be perfectly easy to swell and dissolve a sample of the adhesive in a beaker of solvent but impossible in practice to remove it from a textile or from porous stonework. Does the application of a material alter the information provided by an object? For instance, the temperatures required for joining metals by brazing can alter the microstructure of the object, so an organic adhesive might be preferable. Does the added material produce an unacceptable change in the appearance of the object? A coating may protect metal from corrosion, wood from loss of moisture, and fabric or a painting from environmental degradation, but these advantages have to be weighed up against any resulting difference in appearance and tactile behaviour, as well as the possibility of trapping harmful chemicals in the artefact – salts in stonework trapped by a waterproof barrier, for example.

Decisions about these and many other related questions have to be weighed against the need to join, coat, or consolidate an object in order to prolong its life, to make it easier to handle, to prevent total deterioration, and so on. There are bound to be compromises – the strongest adhesive may be the most difficult to reverse and indeed less likely to break in the event of bad luck and accident than the object itself; the surface coating least permeable to pollutant gases may give too much gloss to silverware. But, again, appreciating the underlying science of the processes involved should help you in your decision-making.

This book is *not* intended to be a practical handbook, but it aims to provide you with the scientific basis you need in order to read other specialist literature more effectively. It will not tell you, for instance, which adhesive to use for a particular repair, but by increasing your awareness and critical faculties it should help you to make a wise decision. Where possible, it will discuss chemical classes of material rather than current products – although a few very well-known products such as Paraloid® B-72 will be named quite frequently. This acrylic resin is used as adhesive, coating and consolidant by conservators working on both organic and inorganic materials, and it is very likely the first example of such that would spring to mind.

This book is divided into three parts, joining, coating, and consolidation, and they are considered in that order. A number of scientific viewpoints are relevant to all three processes, but to maintain a consistent and coherent development of the common topics (such as the chemistry of polymers and the mechanical behaviour of solids) they are all introduced in the context of joining and dealt with there as fully as is needed for consideration of both joining and coating. Structuring the book in this way means that the first part is by far the longest, but this does not mean that joining has been given a more comprehensive or deeper treatment.

However, it does mean that if you are not familiar with the science covered in the first chapters, you are unlikely to understand Chapters 6 and 7. As with the other books in the series, the best way to use the book is to start at the beginning and work slowly through it.

In one sense, there is less to update in this book than for *An Introduction to Materials and Chemistry* in the series. Quite a lot of the text is unchanged, or has been augmented only a little. None of the basic scientific principles involved has become *less* important since the book was first written by a team of conservation scientists and educators in the 1980s. No new, applicable scientific principles have been discovered since then. But the processes of applying coatings and consolidants are more diverse and nuanced: the original text read in 2024 as if it assumed that most objects coming to a conservation department were archaeological ones made from inorganic materials that had been 'excavated' from the nearest museum store.

The original book was written in a period when scientific assessment of conservation treatments had been taking place for only a decade or two, and more so in some countries than others. This was a period when industrial materials only available from mid-twentieth century were being used in preference to the natural materials that had previously dominated in conservation interventions of all kinds. Many of these were derived from animals (gelatine and shellac as adhesives, beeswax for lining canvas paintings for example) or plants (wheatstarch). Now in the 2020s it is possible to look back at a far wider range of possible materials used as adhesives, consolidants, and coatings; to evaluate and compare their performance in real-life museum, archive, and site conditions; and to look forward to greener, more sustainable, and new materials that are appropriate to under-resourced countries that are rich in cultural heritage and also have abundant natural plant-based resources to feed into their own industrial production, and are appropriate to tropical or extreme climates as well as temperate ones. Worldwide, the developments have been in application of coatings and consolidants to challenging, previously untreatable objects such as those with very friable and underbound paint. The actual materials that conservators use in the next decades will probably be pretty different from most of those used today, which are based largely on petrochemicals, but the scientific principles underlying their use, and conservators' aims to preserving objects ethically and with due regard to their significance and future use, will not change.

This updated edition also includes suggestions for further reading, largely drawn from the papers published in recent decades in the leading conservation, conservation science, and heritage science journals, and the fairly small number of recently published books that cover the same topics as this book. This selection makes no claim to be fully comprehensive, as a doctoral thesis might aim to be, but the literature search revealed some interesting trends. The 1990s and 2000s saw studies of the fundamental properties of both the materials found in objects and the materials used to conserve them; some of these conservation materials already being out of production or falling out of favour. The 2000s and 2010s produced review papers on classes of materials used for adhesives, coatings, and consolidants, with summaries about their performance as well as their properties. Some of these papers are in the list for further reading, although many would be of interest mainly to a conservator working on that type of material. Recent papers from the 2020s are nearly all on the topics of new materials, greener materials, or nanomaterials and the improvements they may offer. Interestingly, a clear majority of these papers discuss consolidation of the type of surfaces that have always been the most challenging to the conservator: friable, underbound ones; decaying stone, unfired brick, or plaster surfaces in buildings that are by definition sited in variable environments; and by implication newly excavated material moving from a burial environment to a museum environment. There are fewer papers on coatings or varnishes, and not many whose main topic is using adhesives.

This book has also been written on the assumption that you are familiar with the scientific ideas and vocabulary already covered in *An Introduction to Materials and Chemistry* and *Cleaning.* In particular, you will need to be familiar with the following subjects:

- solubility, solutes, and solvents
- polar and non-polar molecules
- secondary bonds between molecules; dipole-dipole; Van der Waals; hydrogen bonds
- the types and formulae of common organic compounds, particularly alkanes, alkenes, alcohols, esters, and ketones
- surface tension of liquids; how detergents work; the wetting of surfaces; hydrophilic and hydrophobic molecules
- oxidation and reduction

Simple demonstrations have been included to illustrate or clarify certain points in the text. New scientific words and terms are printed in bold type and are also listed in the index. Cross-references are given where it may help you to refer back to a previous passage (or to the other books in the series).

Foreword to the *Science for Conservators* series

The series was first published in 1982 as a series of three books aimed at conservation students and conservators who had come into the field without any scientific training beyond their high school years. It was intended to develop into some six or seven topics covering a broader spectrum of knowledge than the first three, which concentrated greatly on chemistry. The series of three was reprinted in 1987 and again in 1992. It has been in steady demand ever since, despite the text being unchanged since the early 1980s. Some conservation training programmes in a range of countries have recommended the series for pre-programme study, while others have included it in their first year reading list for the students from a humanities background, or for all their students.

Having no training in the natural sciences after the early teenage years, and sometimes even fewer encounters with the biological sciences, had been the background of most conservators in all countries for much of the twentieth century. It still is for some pre-programme conservation students from a few countries, for those who undertake apprentice-style practical training straight after high school, and for others who enter the field of cultural heritage conservation through post-graduate training that was preceded by humanities-based or social sciences–based graduate study. The conservation of artefacts is a truly inter-disciplinary activity which would best be begun with the knowledge of a lifetime! Most conservation training programmes today, including those for heritage scientists, aim to provide pathways into thought processes and knowledge bases from the natural sciences, the humanities generally, and to a lesser extent the social sciences, engineering, the biological sciences, and philosophy and ethics as well, through targeted teaching in all these areas.

The fully revised and updated *Science for Conservators* series is building on this concept. It is intended to serve as a basic primer, and an aid to the assimilation of concepts that will become embedded in the thought processes involved in decision-making by conservation professionals. The series is targeted at several groups: those about to enter a conservation programme without personally having a scientific background; conservation professionals in training and with a similar background; those who are establishing themselves in their challenging and rewarding field; professional conservators who may want to check that they are using a scientific term to mean the same thing a scientist would assume it to mean, rather than its meaning(s) in everyday conversation; and those who have studied the natural sciences or even the conservation of artefacts in another language than English, but are now working, writing reports and publishing in English.

When reading each book, allow yourself to become completely familiar with a section and confident about its contents before moving on to the next. Do not read large portions at any one sitting. Although the series is an elementary one, you will need to take plenty of time in working all the way through it. You should not feel disheartened if your progress at times seems slow. If you do have particular difficulty with a section, ask a conservation scientist, a heritage scientist, or another conservator about it. It is not worth struggling on your own; even a scientist with no knowledge of conservation can help, although one associated with a conservation programme or a museum would be even better. Very often the problem seems surprisingly simple to clear away if you can go through it with somebody else.

Worked examples and exercises have been included where they will be useful. Check your answers at the end of the book. Occasionally, some simple demonstrations are suggested to illustrate or clarify the written text. At relevant points you will also find reference tables. When scientific terms appear or are defined for the first time, they are **emphasised thus** for easy reference. A full index of such terms is included at the end of each book.

Knowledge in the conservation profession is always growing. New journals are established, and the longer-running journals can now be accessed online right back to their first volume. Conferences and events take place every year and are now preserved as recordings of the presentations and discussion sessions as well as in published conference proceedings. There are hundreds of books now written by conservation professionals for their fellow professionals, with much less need for conservators to rely on textbooks written for pure scientists. I am planning that this series will be expanded beyond the three books originally published, to cover a broad spectrum of topics relevant to the conservation of artefacts – just like the original concept of the (short) 1982 series. Further books now include *Colour, Colour Measurement and Colour Change* and *Experimental Design and Scientific Data Analysis*.

Knowledge of health and safety has changed and developed a great deal since the 1980s, of course. The conservation profession has also evaluated the effects of past treatments on the preservation, appearance, and indeed survival of objects, and was beginning to do this, and publish such evaluations, in the 1980s. Many treatments that were common before and in the 1980s, and some chemicals and reagents that were then available on the bench in most conservation workspaces and every undergraduate chemistry laboratory, now appear hazardous to objects, dangerous to the user and inappropriate to use, in the third decade of the twenty-first century. Yet they played a part in the long history of a great many artefacts: it is necessary to know what might have been done to an object in the past in order to treat and sometimes stabilise it now. Also, treatment records used the terminology of their era. I have taken the decision not to exclude from the updated series the materials that are no longer used by responsible conservation professionals. They were used at the time for their unique properties and perceived efficacy, and they may have affected the properties of an object and its likelihood (or not) of surviving for another few centuries. The reasons why they fell out of regular use will however be mentioned.

Some materials derived from petrochemicals are going to be much less used in the future, as greener alternatives are developed and come into wider use. Green alternatives made locally will become cheaper than imported chemicals, or those made from petrochemicals. Using greener materials and cutting down on both waste and transport are very active topics of discussion in the field of conservation at present, and sustainability aligns with the profession's core values too.

1 Sticking things together

In this chapter the question 'What makes things stick together?' is tackled. The requirements of a good adhesive and a good joint are also considered, and then 'service conditions': the environment and degree of handling that the object is expected to meet once it leaves the conservation studio. Finally discussed are the characteristics, advantages, and disadvantages of the various types of adhesives used in conservation.

A How do things stick together?

The objects you deal with are integral entities, that is, they stay in one piece, because they are held together by the strong cohesive forces that stem from either ionic, covalent, or metallic primary bonds between their constituent atoms (discussed in *An Introduction to Materials and Chemistry*, Chapter 4) or from secondary bonds between their molecules. As you know from common experience of a broken object, a cup say, even if you fit the pieces back together very carefully they will not stick by themselves (or at least, not enough to withstand even gentle handling), and you can still see the line of the break. Although the thickness of this line can be extremely small – a fraction of a millimetre – it is still very large compared with the distance over which the **cohesive forces** act.

One reason why the broken surfaces cannot be put back into intimate contact is that on a microscopic scale they are, in fact, very irregular. Even the fracture surface of a piece of glass, which we see with the naked eye as being extremely smooth, is revealed by an electron microscope to be a mazy scene of ridges and valleys (Figure 1.1). There were probably micro-losses when the fracture occurred. It is not surprising therefore that two such surfaces cannot be perfectly matched again: they will only make contact on a microscopic scale at rare points, and over most of the surface there will be an air gap between the pieces.

Further, immediately after an object is broken, the newly created fracture surfaces become **contaminated** by oxygen, water, or other chemicals in the environment. Those atoms or molecules at the surface of the object that, prior to fracture, contributed to the cohesive forces in the material are now free to make bonds with the contaminant atoms or molecules. In this way metals oxidise by primary bonds being formed between metal and oxygen atoms; water molecules from the

DOI: 10.4324/9781003389217-1

(a)

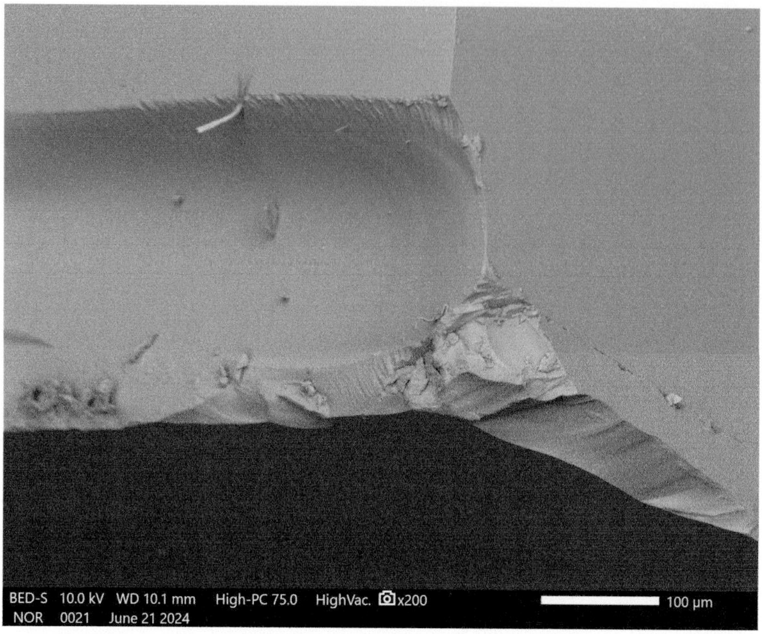

(b)

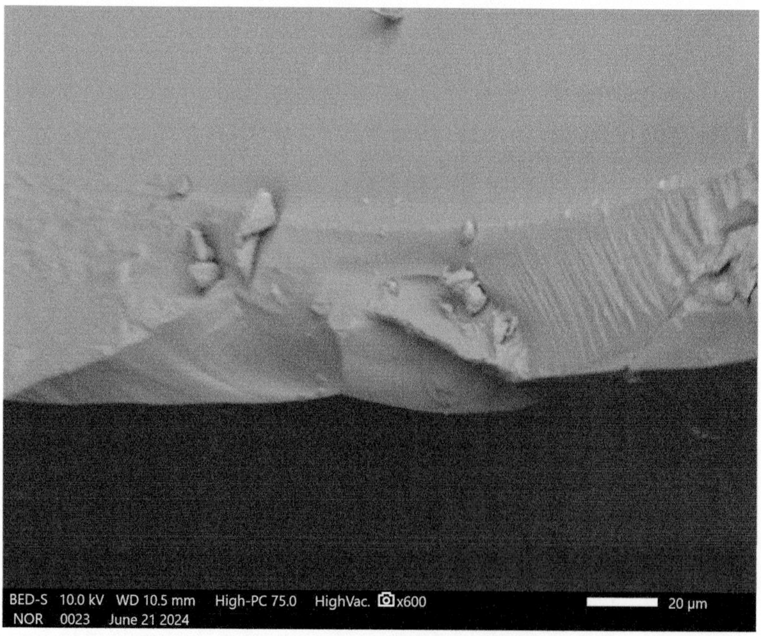

Figure 1.1 Backscattered electron images of newly broken and clean soda glass, coated with 80:20 gold:palladium. (a) Break imaged at ×200; (b) fracture edge right of centre of (a) imaged at ×600.

Images: Joyce H. Townsend with Innes Clatworthy, Natural History Museum Electron Microscope Unit.

atmosphere hydrogen-bond to materials such as paper and wood leading to a thin layer of moisture on the surface, and so on. Processes like these are the ones that have to be reversed when an object is cleaned. Surface contaminants *adhere*, via primary or secondary bonds, to a fracture surface, and so they inhibit the rejoining of the broken pieces. The 'ideal' fragments to be adhered together would be (returned to) the same condition as the instant they were fractured – but that is impossible to achieve. And over time – and especially burial in the archaeological context – the fragments might be abraded and suffer further losses.

Since the broken surfaces will not join themselves together, another way must be found. Basically, there are two ways by which bits of solid materials can be joined together:

- Using some sort of device that mechanically locks the pieces together. (Mechanical in this context means non-chemical. It may be a temporary means such as a box of fine, clean sand or other powdery material to support the pieces until the adhesive has time to set and gain its full strength, or low-tack tape holding them in place until the adhesive sets.)
- Using an **adhesive**; that is, a material which, ideally, fills the gaps between the pieces, adheres to both surfaces and achieves a sufficiently strong and rigid interface between the pieces to support its weight and render it suitable for display and interpretation, although not necessarily for the function for which it was made.

Sewing torn pieces of fabric together is an obvious example of mechanical locking, and so is the protection of delicate fabric such as deteriorating silk by covering it with a thin, sheer, colour-matched fabric and couching it down with tiny stitches made with a fine sewing needle on the viewing side, the thread carried along the back of the original until it is sewn through again at another point of contact between the two fabrics. More typical mechanical locking involves the use of screws, rivets, and/or dowels. Historically, fine screws were used on occasion to fix together fired ceramic objects such as broken porcelain, with the loss of original material where a hole was drilled out for the screw. Riveting is a traditional construction method for joining thick sheets of metal, extensively used in ship building and bridge building, which provides a very strong joint. Dowels or armatures in cast metal sculptures might be part of the original fabrication method, or a necessary repair of a fracture, to help a sculpture support its sometimes very considerable weight. The joining mechanism does not depend on any primary or secondary bonding between the locking device and the pieces of the object. Of course, chemical reactions may occur over time; the rusting of screws with migration of rust stains into the ceramic body, the rusting of dowels in outdoor sculpture until the dowels suffer sudden fracture, or the rotting of threads used to couch down a textile repair are examples.

Adhesives have been the subject of considerable scientific research, but exactly how they bond to all solids is still not fully understood. It is thought that adhesion is due to secondary bonding between the molecules in the adhesive and the atoms or molecules at the surface of the pieces to be joined (the **adherend** or **substrate**). Clearly, the magnitude of these forces is an important factor in determining the strength of the resulting joint. Some adhesives are formed by chemical reactions *in situ* between different ingredients. This happens, for instance, with epoxy resins, polyester resins, and cyanoacrylate adhesives. Of course, in a chemical reaction, primary bonds in the starting materials are broken and new ones are made to form the products of the reaction. Even with these adhesives it is thought that no new primary bonds are formed across the interface between the adhesive and the solid surface, but that adhesion is essentially by means of secondary bonds.

In some particular cases the joining process does produce primary bonding right across the interface, so much so that the original interface disappears. This happens, for instance, when a joint is made by locally melting the two pieces and holding them together until they resolidify; examples are brazing and welding, and also riveting, all of which are carried out at elevated temperatures. These last examples are designed to alter the local microstructure of the metal permanently in order to confer strength comparable to that of a larger piece of the same metal with no join present.

The difference between the two ways of joining – by mechanical locking or by an adhesive – becomes more fuzzy when you consider how an adhesive works at a microscopic level. In fact, in many cases, **mechanical interlocking** may contribute significantly to the effectiveness of an adhesive. If you think about a microscopically rough fracture surface, you will appreciate that a suitable adhesive could penetrate the nooks and crannies on the surface, and also the surface to which it is to be joined, and thus produce an interlocking of the fragments, as well as filling in the voids left by micro-losses. Similarly, pieces of paper or fabric can be joined by an adhesive that penetrates between the fibres making up the threads, as well as into the weave structure. The pieces are held together by the veins of adhesive, which act as hooks around the paper or textile fibres. Historically, artists and map-makers joined paper to form larger sheets this way, and the wrong adhesive used to adhere a print or drawing to a secondary support could only be reversed with an undesirable degree of mechanical action, leaving torn or ‘skinned’ paper afterwards.

In the context of this book there is no need to pursue the details of theories of adhesion any further, however the basic mechanisms of secondary bonding and microscopic interlocking just described will enlighten later discussions. In the next section the characteristics that an adhesive and the surfaces to be joined should have in order to produce a satisfactory joint are explored.

B What is needed to make a good joint?

What is meant by a good joint really depends on the meaning of ‘good’. It involves a balance between judgements about the quality of the joint on technical, aesthetic, and ethical grounds, the required properties of the adhesive and the efficiency with which the joint is made.

At this point it is important to appreciate that in discussing different types of adhesive, and the science relating to them, there is a very particular definition of ‘good’ when considering adhesives for conservation work. Most modern adhesives have been developed to produce joints that are good in terms of the criteria of mass-production manufacturing industries – furniture production, car production, and so on. This has meant that the search has been for adhesives that produce very strong joints – indeed with many modern adhesives the material of which the object is made may break, rather than the joint. Further, the requirement is for joints which resist degradation in use caused by chemicals in the environment. This usually means that joints produced by such adhesives are difficult to reverse by, for example, dissolving in common solvents such as water and simple organic liquids. These needs for strong, chemically indestructible adhesives are usually at odds with those of the conservator, who often wishes to make a glued joint which holds the pieces of an object together satisfactorily but which can be readily taken apart at some future time, by a later generation of conservators, without damaging the object.

Although these very different requirements are quite obvious, ways of considering the scientific background to adhesives in the context of conservation are *not*. For instance, it is more straightforward to analyse the factors that produce a strong joint rather than those that produce a weak joint. This is the approach taken here, with, where appropriate, a discussion of the special requirements of adhesives in conservation work.

Think of an object that needs to be mended. What are the important requirements of the joint you intend to make? What are the steps in making the joint that need the most careful attention?

The flow chart in Figure 1.2 attempts to cover the main points you should have considered. Notice that many of the factors have implications for others. For example, making a joint strong may not be compatible with the necessity to remove (take down) the joint without harming the object decades later, and an invisible join in colourless glass with a transparent adhesive may become visible because it turns yellow, then brown. All of the topics in Figure 1.2 (except the need to have clean surfaces) have implications for the type of adhesive you use; they can be used to

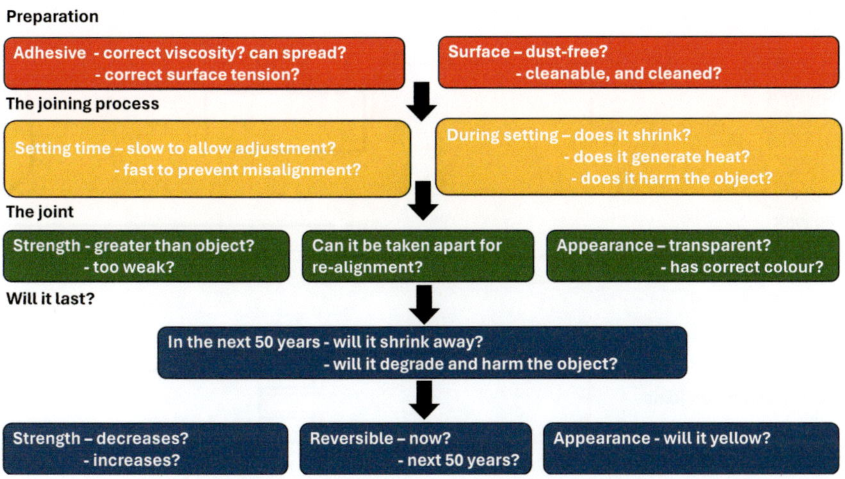

Figure 1.2 Factors to consider when selecting an adhesive. A flow chart shows the factors to be considered in turn, when the type of adhesive is being decided.

draw up a specification of the desirable properties of an adhesive. However, there may be additional constraints put on the choice of adhesive, such as cost, availability during the excavation season, and whether it presents health hazards to the conservator, for example by releasing solvent vapour as it cures, or hazards to the object in the future as the adhesive degrades.

You will now be aware that the selection of the best adhesive is a complex problem. Of course, for a particular job some of these factors may be irrelevant or easy to decide, but, even so, a large number of variable and inter-related factors will remain. Having considered the problem in a general way, you can now start to analyse the main factors in more detail, using the items in Figure 1.2 as a guide. A convenient starting point is with the need for the adhesive to be spread over the whole of the surfaces to be joined.

C Types of joint

Many breaks that have to be repaired require edge-to-edge joining, an arrangement which is called a **butt joint** (Figure 1.3 top). A joint like this, in isolation, is in danger of breaking if the pieces are flexed or rotated, so the adhesive should have a similar strength to the object, rather than being weaker. If the fragments being joined combine together to make a rounded shape like a vase with no missing pieces, flexing is not a problem once the whole object has been assembled, glued, and set: the whole assembly has more rigidity due to its own mechanical interlocking. An adhesive weaker than the material of the object would be a good choice, in case of impact, falling, or severe shaking, because then a repaired break might fail rather than a new one occur. Overlapping joins known as a **lap joint** are less common in conservation practice, if a break is being repaired, although strip-lining a canvas or adding fabric to suspend a tapestry under even tension are examples. Figure 1.3 (middle) shows the simplest possible overlap, but when this is used in

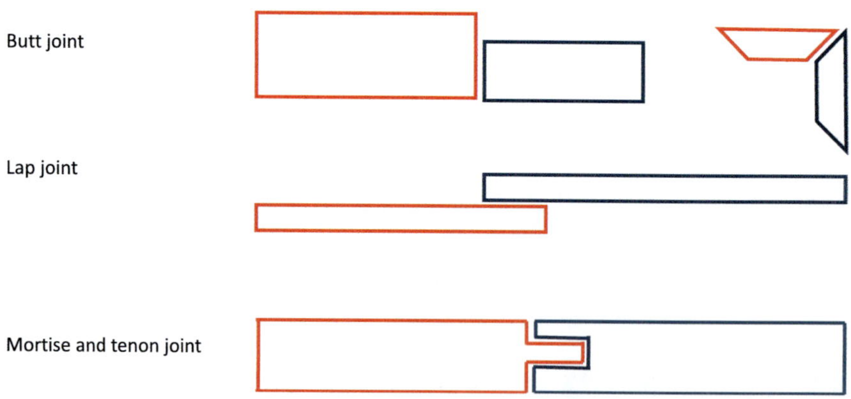

Figure 1.3 Top: two simple forms of butt joint. Middle: a simple lap joint. Bottom: a simple form of socket joint made as a mortise (on the left) and tenon joint with every cut forming a right angle. (Which means it is not a dovetail joint.)

construction, the two pieces usually have a more complicated shape where they meet, to form a partly interlocking joint. When inside edges are angled, they may form a **dovetail joint**. Joins with sockets, where one piece is inserted into another to form an interlocking joint are more common: an example would be the insertion of each step for a staircase into a corresponding hole cut into the supporting wall. The slightly more complicated joining process shown in Figure 1.3 (bottom) is of the type called a **mortise and tenon** joint, also spelled sometimes as mortice and tenon. This is the simplest example, and more complicated designs might be used by cabinet makers and frame makers to confer greater strength. Wooden doors, furniture, and stretchers for canvas, or the steps for a self-supporting staircase, were often constructed in this way historically, with the two shaped pieces glued with a (usually slow-setting) strong adhesive as well as cut to fit together, with or without a dovetail. Breakage of the material rather than failure of the adhesive is the most likely result of an accident here, and such joints are not meant to be taken apart easily.

D Service conditions

This covers the environment that the object and its adhesive will encounter post-treatment. The manufacturer's data for industrial adhesives and coatings will usually include information on the temperature range over which the product will perform well, and perhaps a pressure range too. It might be stated that the product must be stored in a frost-free environment, or kept in a cool, dry place before use. In museum conditions in temperate climates, conditions are so benign that it is possible to forget that adhesives can soften or even melt at higher temperatures. This is of practical importance in tropical climates, especially for objects such as outdoor sculptures in any climatic zone that may receive direct sunlight – the surface temperature of a dark surface could reach 70°C in such circumstances. (Because a dark surface can be heated around 20°C above air temperature in the shade.) Widely used acrylic adhesives such as Paraloid® B-72 exposed even to 50°C would be so soft at that temperature that flow of liquid adhesive out of the repair would not be surprising. Above 40–45°C there would be some softening, and that is not an impossible temperature for a modern building with no passive or active environmental control to reach in Mediterranean countries as well as in continental climates at low latitudes, not to mention tropical climates.

Some adhesives might become brittle at low temperatures approaching freezing, causing joints to fail due to low impacts that would not cause damage at room temperature range: acrylic paints, for example, are more brittle in these circumstances, and so might acrylic adhesives be.

Another environment that might affect a conservation adhesive more than a wall painting made in appropriate materials for the climate is the constant presence of moisture from rising damp, or the occasional presence of condensation within a poorly ventilated building in winter or during the monsoon season. This could cause thin film residues of a normally transparent and colourless adhesive to grow opaque. Any moisture penetration below the residue or into the joint might lead to separation of adhesive and object, and failure of the joint.

E Helping the adhesive to do its job

Whatever the mechanism of adhesion it is self-evident that the greater the area of intimate contact between an adhesive and the object, the more effectively will the surfaces be stuck together. This is simply because the greater the area of contact, the more bonding sites there are. To obtain a maximum contact area, three things are important:

- The surfaces must be as clean as possible, that is as free as possible from chemicals that obstruct bonding between the adhesive and the adherend. Any dirt, friable corrosion such as rust, or loose pieces of the object not firmly attached to the surfaces will prevent contact by the adhesive. But their thorough removal might be damaging to the object.
- The rougher the surfaces are the better, because the surface area is greater than for a smooth surface and because interlocking of the surface and the adhesive is increased. This is why the instructions on the containers for many proprietary adhesives tell you to roughen the surfaces to be joined. Such treatment is not appropriate for most conservation work, but it may be useful to know that roughening materials such as wood and leather raises some of their fibres proud of the surface, thus giving enhanced bonding by embedding the fibres in the adhesive layer.
- To achieve close contact and spread over the surfaces, the adhesive must wet the surfaces, be liquid enough to flow into pores and crevices in the surfaces, but not so free-flowing that it will pool in microscopic hollows and leave higher points adhesive-free.

Whether a liquid can wet a solid depends on its **surface tension**. The lower the surface tension of a liquid the easier it is for a droplet to spread and wet the surface of a solid. If there are strong forces of attraction between molecules in the liquid and the solid, the liquid will wet the solid surface. If, however, the secondary bonding within the liquid is stronger than the bonding between liquid and solid, then droplets will be formed that do not spread to cover the surface. (You can demonstrate this by dropping water onto a cleaned (degreased) sheet of glass and onto a piece of glass that has been smeared with candle wax, or one that is well covered with fingerprints.) As the sticking power of adhesives depends on secondary bonding to the solid surface, the majority of adhesives have strongly **polar groups** in their molecules. If the surfaces to be joined are contaminated with non-polar substances such as grease, the adhesive will not spread.

So far only flat surfaces have been considered. The extent to which a liquid penetrates an irregular surface depends on another phenomenon related to surface tension: **capillary action**. You take advantage of this effect whenever you use absorbent materials to soak up liquids – paper tissues and towels for instance. These materials consist of a network of fibres, and liquids soak into them by being drawn along the narrow interstices between the fibres by capillary action. The same effect occurs in thin tubes when they are dipped in most liquids; the liquid rises up the tube. Such tubes are called capillaries. The pores, cracks, scratches, or pits on the surface of an object can also act like capillaries.

For capillary action to take place the liquid must wet the surface of the solid; the adhesion between the liquid and the solid is then greater than the cohesion of the liquid. This means that the liquid/air interface (the **meniscus**, which takes the Greek plural menisci) is curved downwards because the (secondary) forces of adhesion between the liquid and the material of the tube pull the liquid up the sides of the solid surfaces. This is illustrated for capillary tubes in Figure 1.4. The liquid rises until the upward pull is balanced by the gravitational pull on the liquid in the tube, Figure 1.4 (left). Another important feature of capillary action is that the wider the capillary, the smaller is the capillary rise, as in Figure 1.4 (right). This follows from the balance between the upward force and the weight of liquid in the tube. A tall, narrow measuring cylinder that holds 10 ml or 20 ml will show a visible meniscus when it is filled with water, which has a high surface tension. Most common solvents show this effect less, and a few drops of a solvent like ethanol that mixes with water will flatten the meniscus a lot.

This is not the end of the story, however. In order to flow over a surface, a liquid has to be free-flowing, like runny honey, although a bit less flow would be easier to work with at typical studio temperatures, whilst assembling several fragments into position for a repair. The resistance a liquid offers to flowing is called its **viscosity**. In general, the larger the molecules in a liquid the stronger the secondary bonding

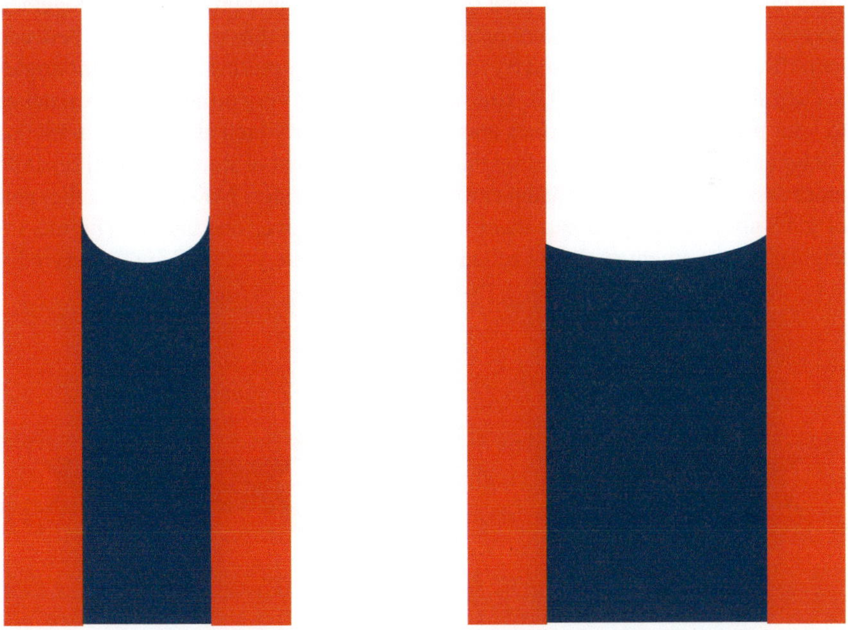

Figure 1.4 Schematic of liquid with high surface tension (blue) in a narrow capillary tube (left) and a wider one (right). The meniscus is flatter in the wider tube.

between them, and the greater is the viscosity. The lower the viscosity, the more readily will a liquid spread. From everyday experience, you know that the viscosities of solids are extremely high compared with liquids: solids do not flow readily, over the timescale that we are watching them in person. But very viscous and tarry materials do flow over centuries,[1] which is the timescale of the lifetime of objects in museums. One might expect that very old repairs with such materials could be discovered, to stone building perhaps, but this author is not aware of any that have been reported.

You can probably see the importance of the need for an adhesive to wet the surfaces of the adherend, that is, to possess **wettability**, and to have a low viscosity. The adhesive then makes intimate contact with the surfaces and penetrates the irregularities on them. Up to a certain size, the wider the pores, hollows, and so on, the less they will be penetrated. It clearly follows that, at this stage, the adhesive has to be in the form of a liquid. But intuitively you know that finally the adhesive must become a solid.

Demonstration

Obtain four ordinary glass microscope slides and carefully clean one face of each by washing them in acetone and leaving them to dry. Being careful not to touch the cleaned faces (grease from your fingers will destroy the effect you are seeking), put a few drops of water on the cleaned surfaces of two of the slides. Place one of the other cleaned slides on top of each of the wetted ones and press the pairs together. You should now have two pairs of slides, each with a thin film of water between them. Take one of the pairs and try to separate them by pulling them at right-angles to each slide (Figure 1.5 (left)). You will find that it is very difficult. The resistance to your pull is due to the surface tension of the water which attempts to minimise the liquid/air interface. If you drop the slides into a bowl of water they will separate more easily since surface tension exists only at the surface. Now take the other pair of glass slides and try to part them by sliding them over one another (Figure 1.5 (right)), a movement which is called **shear**. This time you will find it very easy

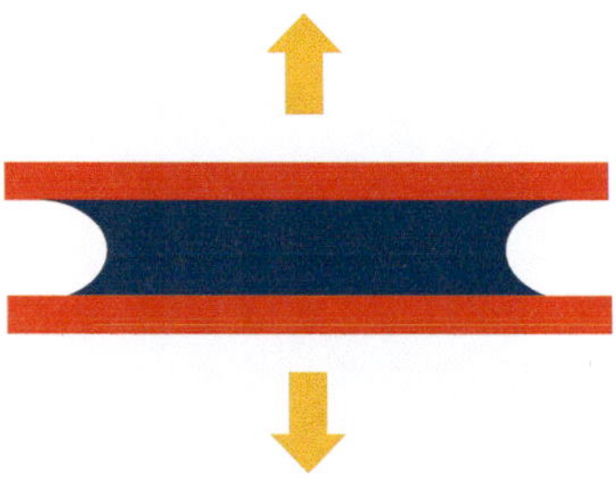

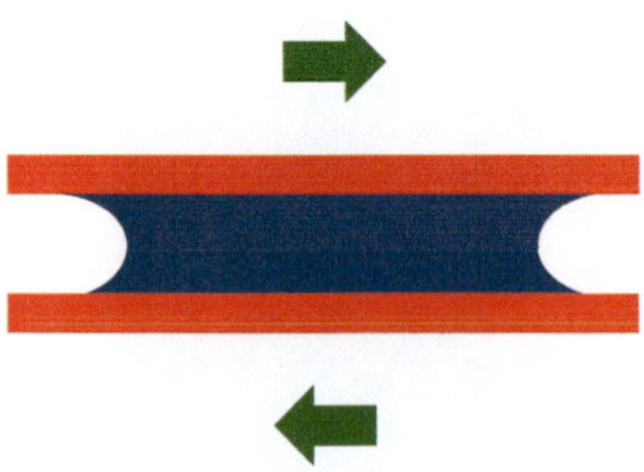

Figure 1.5 Schematic of liquid with high surface tension (blue) trapped between two microscope slides that are (left) being pulled apart and (right) being slid apart. The sliding apart is the application of a shear force.

to part the slides. In this case, surface tension does not provide resistance to your shearing action because, as you will notice, the two menisci remain intact as the slides move past one another. The water continues to adhere to the glass, so the shearing movement must occur within the film of water. This is because the water molecules can move past one another very easily. (The ability to flow fairly readily is one of the characteristics that distinguishes a liquid from a solid.) If the water were frozen, both types of separation would be hard to achieve, and the glass slides might well be broken first. (Best not try this!)

The background to one of the fundamental problems of adhesives has now been outlined: to wet the surfaces, the adhesive must be a liquid, and to form a strong joint it needs to be a solid. In practice, the adhesive is applied in the form of a liquid and then modified to produce a solid. Basically, there are three ways of doing this.

- **Melt-freeze adhesives**. The adhesive is prepared as a solid, is heated and applied to the surfaces to be joined as a molten liquid, then allowed to freeze back to a solid. Examples of this type of adhesive used in conservation are waxes such as beeswax, some animal glues (both of these survive the gentle heating with no loss of performance), soldering, and brazing. Using a hot glue gun is another example – although this convenient tool is best used only for temporary packaging and cushioning for an object, and not for conserving it.
- **Solution adhesives**. Here, the intended adhesive material is dissolved in a suitable solvent or, in the case of proprietary adhesives, a solvent mixture. After the solution has been applied to the surfaces, the solvent evaporates, leaving the solid adhesive behind. Examples of this type are starch-based glues, shellac (dissolved in alcohol), gum arabic (dissolved in water), rubber solution (for repairing bicycle inner-tubes), and polyvinyl alcohol (PVOH). A variant of this type is to have the adhesive as a powder that is mixed with a suitable liquid to produce a **dispersion**. (Emulsions, slurries, and pastes are forms of a dispersion, although an emulsion consists of one liquid dispersed in another liquid.) Again, the liquid evaporates to leave the solid. An example in conservation is polyvinyl acetate emulsion (PVAc). (Note that the abbreviation 'PVA' has been used for either of these, in some publications on case studies of treatments, which is unclear. 'PVA paints' in fact have a medium of polyvinyl acetate.)
- **Reaction adhesives**. In this type, two or more liquids are mixed and then applied to the surfaces to be joined; the chemicals react to produce a solid adhesive and rapid application may be necessary. Common modern examples of this type are based on epoxy resins and styrene-based polyester resins. (Both of these harmful materials should be used only with suitable protective equipment such as disposable gloves and fume extraction.)

Many adhesives are synthetic polymers of various types. Many of the traditional adhesives, which you may come across in objects that have been repaired previously even if they are not so commonly used today, are based on naturally occurring polymers such as starch, cellulose, rubber (latex), and proteins obtained from animals (gelatine, shellac) or plants (wheatstarch). In order to continue this analysis of factors involved in making a good joint, you need to be familiar with some of the scientific models used to interpret the structure and behaviour of polymeric materials. This is the subject of the next chapter.

Note

1 See https://en.wikipedia.org/wiki/Pitch_drop_experiment for examples.

2 The chemistry of polymers

Polymer is the name given to very large molecules which contain polymer hundreds, thousands, or even millions of atoms and are made up by the successive linking together of one or more types of small molecule. The small molecules are called **monomers** (from two Greek words, meaning single part) and they produce a polymer (many parts, where 'many' is a number in the thousands or sometimes hundreds of thousands). Most of the properties of a polymer are strongly influenced by the way in which it is built up from its component monomers, and to help draw out this point it is useful to classify polymers into two types: those which form **long-chain molecules** and those which form one large, continuous, interlocked molecule consisting of a two-or three-dimensional network, that is, **network polymers**. In this chapter we will look at each of these in turn.

A Long-chain polymers

These have the simplest structure and can be envisaged as a series of building blocks in which the monomers are each a single unique block with an unfilled chemical bond at each 'end', so enabling the long-chain polymer to be built up by linking one block to another in a growing chain (Figure 2.1).

The simplest polymer like this is **polyethylene**. As its name implies, the monomer is ethylene. It has a double **covalent bond** between the carbon atoms (see *An Introduction to Materials and Chemistry*) which is highly reactive. In effect, one of these bonds provides the stud which links to a neighbouring molecule, which links to the next and so on, to create the long-chain polymer (Figures 2.2–2.3).

The chemical process of joining the monomers is called **polymerisation**. This will be looked at a bit more closely in the next section. Polyethylene is a synthetic polymer which was first produced commercially in the late 1930s. It is the basis of one of the most widely used plastics, polythene, which you meet as plastic bags and all sorts of packaging and containers. In common usage polyethylene is now called **polythene (PE)**. Although the words polymer and plastic tend to be used synonymously, strictly they have different meanings. A polymer is a large molecule; a plastic is a product consisting of a mixture of polymer(s) with a variety of materials added for specific purposes, such as pigments, compounds to inhibit breakdown of the polymer by light or oxidation, **plasticisers** to facilitate shaping and to make

DOI: 10.4324/9781003389217-2

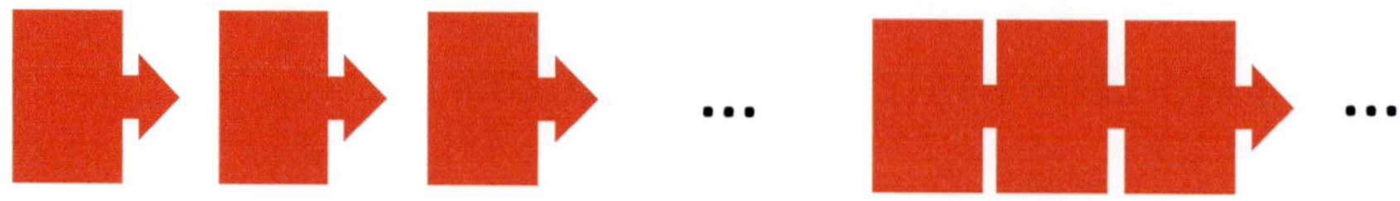

Figure 2.1 Schematic of monomer blocks beginning to join up to form a polymer. The 'open' chemical bond shown by the arrow is still available in the polymer, so more monomers can be joined to it.

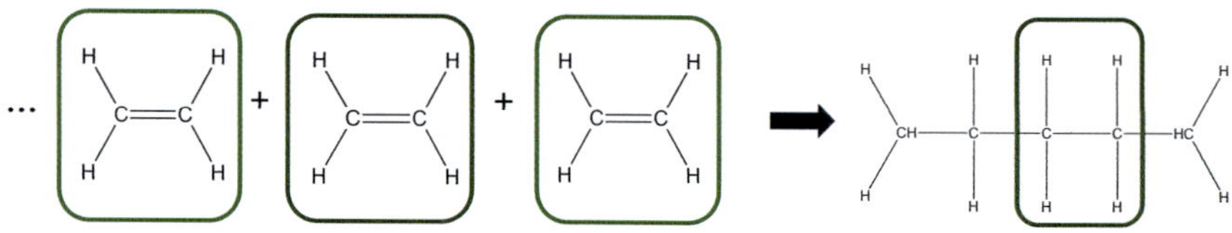

Figure 2.2 Ethylene monomers begin to combine to make polyethylene, a long-chain polymer.

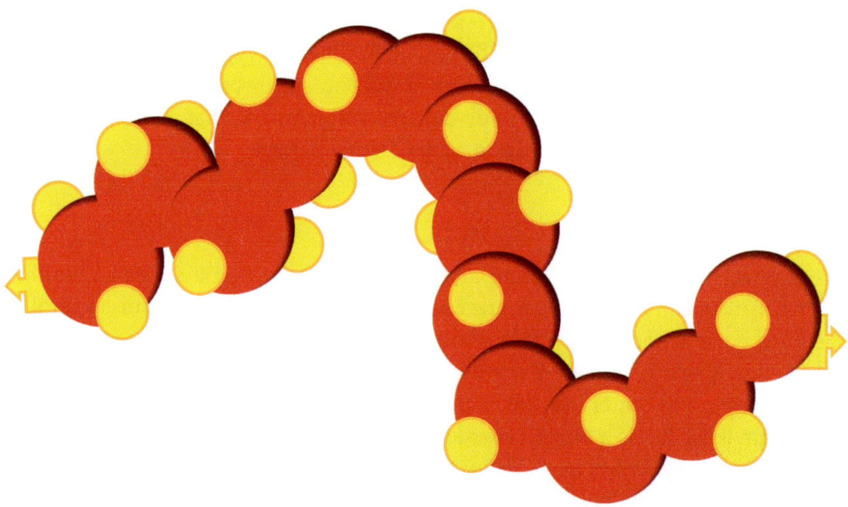

Figure 2.3 Schematic of (a very small) part of a polyethylene polymer, with carbon atoms shown in red and hydrogen atoms in yellow.

the product softer, and so on. Some of these additional constituents can cause ageing problems in adhesives and coatings, and examples of this are dealt with in Chapter 5.

A structural formula is a useful way of showing, schematically, how the atoms are disposed in a covalently bonded molecule. In *An Introduction to Materials and Chemistry* you were warned of the limitations of this type of model. It does not give much idea of what a polymer molecule actually looks like. For instance, a carbon atom is larger than a hydrogen atom. Further, the structural formula model implies that a long-chain polymer is straight and stiff; a real molecule is usually flexible and contorted. Although the C–C covalent bond is strong and fixed in direction, the kinetic energy possessed by the molecule allows rotation of the atoms about these bonds, so the chain is twisted, and in fact it is twisting like a snake moving along rather than being static.

Other long-chain polymers can be described using polyethylene as a basis. As an example, Figure 2.4 depicts the formation of polyvinyl chloride (PVC). The monomer is vinyl chloride.

Vinyl chloride can be visualised as ethylene with a side-group of one hydrogen atom being replaced by a chlorine atom. Many other long-chain polymers can be described in the same way. Figure 2.5 gives two examples that form the basis of important commercial plastics, PVAc (on the left) once commonly used as a conservation adhesive, and poly(methyl methacrylate) (PMMA, Perspex®, Plexiglas®, on the right) which is used as the glazing in frames and to construct display cases and stands, as well as by some sculptors.

You may be perplexed by the names of the polymers listed in Figure 2.5. Some of the difficulty arises from the variety and complexity of the materials themselves and is made worse by the lack of a universally agreed system for naming them. Thus, some are named after part, or all, of their chemical constituents (as are both in Figure 2.5), others have uninformative generic names (for example, acrylic), while some are often referred to by proprietary names. A systematic method of naming is necessary and is provided by the International Union of Pure and Applied Chemistry (IUPAC). Meanwhile, there are standard abbreviations used by conservation professionals for most polymers, such as PMMA for polymethyl methacrylate. Less correctly, these abbreviations are usually known as the names

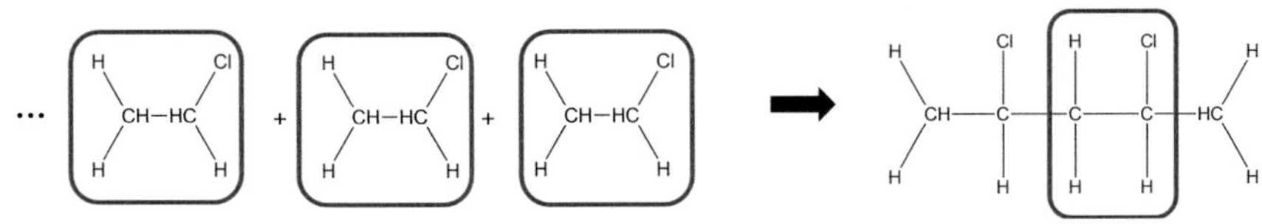

Figure 2.4 Vinyl chloride monomers begin to combine to make polyvinyl chloride, another long-chain polymer.

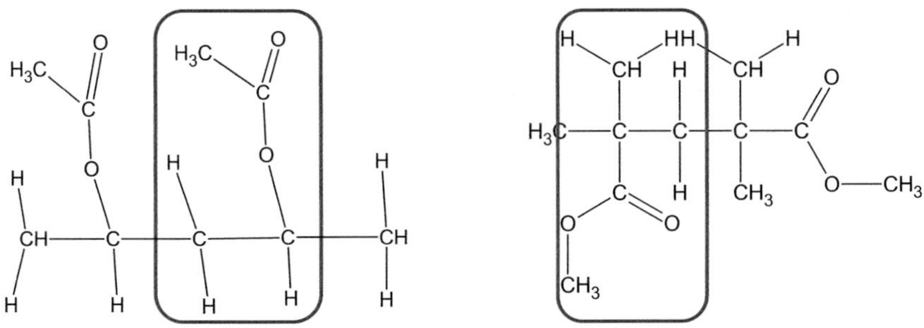

Figure 2.5 Polyvinyl acetate (PVAc) polymer with the monomer indicated by the box (left), and polymethyl methacrylate (PMMA, Perspex®, Plexiglas®) likewise (right).

of plastics, not polymers.[1] You should get into the habit of using these standard abbreviations; it will help, for example, to clear up the confusion about the ambiguous abbreviation 'PVA', which can refer to a product based on **poly(vinyl acetate)** (PVAc) or **poly(vinyl alcohol)** (PVAL or PVAl) as mentioned earlier.

There are many reasons for the commercial development of so many types of synthetic polymeric materials, but one of the most important is that different types of molecule produce different properties in the bulk material. One of the technological aims of their development is to produce polymers that carry out particular functions, or that do them better than alternative materials.

Imagine, for example, a polymer made up of polyethylene molecules and another made of methyl methacrylate molecules. Both materials will contain large numbers of long-chain molecules, but the types of molecule are very different. Ethylene molecules have only hydrogen side-groups attached to the carbon backbone, and, as you know, these are non-polar groups and only very weak secondary forces are set up between neighbouring molecules. As a direct consequence the molecules can be drawn apart quite readily, which means that PE has a low melting point and is relatively soft and flexible. On the other hand, PMMA has some large polar side-groups which means that the molecules are irregular in shape and more strongly bound to one another. Thus, PMMA has a higher melting point than PE and is harder and more rigid.

So far only long-chain polymers that have a backbone composed solely of carbon atoms have been considered. However, the possibility of substituting other atoms for carbon in the backbone provides many other variations. The requirements are that the substitute atoms form at least two covalent bonds with neighbouring atoms so that the chain is maintained. Examples are oxygen, nitrogen, sulphur, and silicon. These molecular structures can be quite complex: Figure 2.6 shows Nylon® 6. There is a range of different nylons, which are distinguished by a number that indicates the number of successive carbon atoms in the backbone of the monomer(s). Some **nylons** are described by a single-digit number – for example, Nylon® 6 which means the polymer was made from a single monomer containing six carbon atoms. Other nylons are described by a larger number – for example, Nylon® 610 which means that the polymer is made from two different monomers, one containing six carbon atoms and another containing ten. Polymers made from more than one kind of monomer are called **co-polymers**.

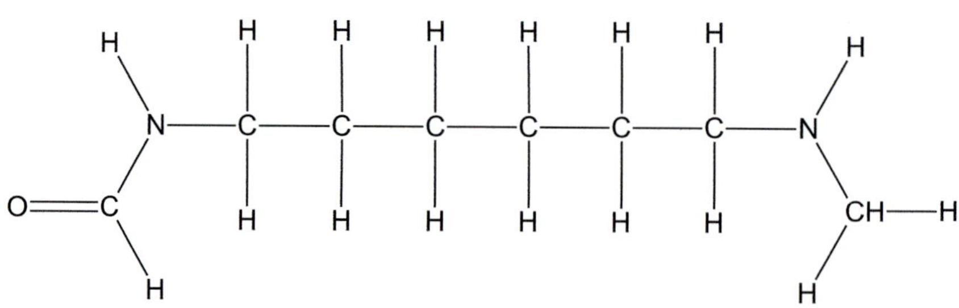

Figure 2.6 Monomer for Nylon® 6.

A number of co-polymers play important roles in conservation work. For example Paraloid® B-72, which is used as an adhesive and coating for many materials and for paint as well, and sometimes as a consolidation agent, is a co-polymer of ethyl methacrylate (EM) and methyl acrylate (MA), which together are abbreviated to EMMA. Here, juggling the relative numbers of the methyl (CH_3–) side-group and the larger ethyl (C_2H_5–) side-group is used as a way of achieving desirable properties, and one proportion has been in use for many years, although the earliest product (long unavailable) had a slightly different proportion of the two monomers. BEVA® 371, an adhesive widely used in the more recent past for lining canvas supports for paintings and weakened textiles for suspending on a wall, and still used for strip-lining, is a co-polymer of ethylene and vinyl acetate together with a variety of other compounds (including harmful aromatic solvents), each of which is added to achieve specific modifications of the properties of the plain co-polymer. (Hence its abbreviation of EVA, preceded here by the name of the paintings conservator who first advocated its use, Gustav Berger.)

The repeated units in a co-polymer can be arranged in different ways, again leading to changes in properties. There are three main types: the **alternating co-polymer**, the **random co-polymer**, and the **block co-polymer**; they are illustrated schematically in Figure 2.7. Co-polymers with at least three basic units are

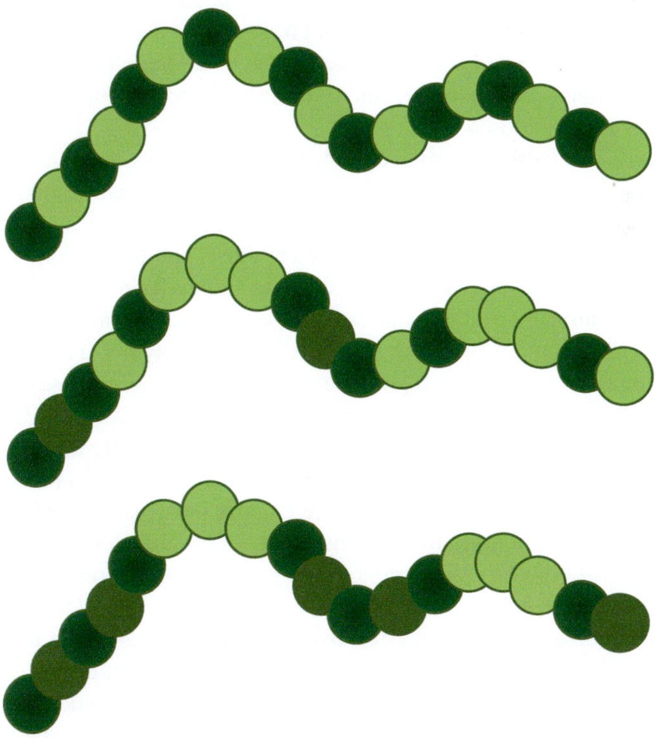

Figure 2.7 Schematic of (top) an alternating co-polymer, (middle) a random co-polymer, and (bottom) a block co-polymer.

made synthetically, and in nature, proteins are co-polymers of up to 21 amino acids. (This is why there is such an enormous variety of proteins; 21 different structures can be combined in a huge number of ways, the simple proteins with fewer than 21 amino acids.) We have now discussed the basic types of long-chain polymers. Most are made synthetically; nature tends to produce network polymers. However, some long-chain polymers do exist in nature, for instance bitumen (a complex mixture of hydrocarbons) and cellulose. Before looking at network polymers it will help if we first consider the mechanisms of making long-chain polymers, that is, of **polymerisation**.

B Polymerisation

Molecules like ethylene and methyl methacrylate can be persuaded by chemical means to join together consecutively using one of the bonds in the double carbon bond, C=C. In principle any monomer with a double bond between carbon atoms can be polymerised in this way; this is called **addition polymerisation**. Notice that the polymer is the sole product of this sort of reaction (apart, of course, from a great deal of heat).

Some long-chain polymers, and network polymers, are produced by a chemical reaction called **condensation polymerisation**. The formation of the peptide link is a typical example of a condensation reaction. In such a reaction two different sorts of molecule react to form the desired product as well as a small molecule as a by-product, which is usually but not always water. The long-chain polymers known generically as nylon and polyester, and also cellulose and proteins, are polymerised by successive condensation reactions.

Since cellulose-based adhesives are often used by conservators, the formation of cellulose will be considered in some detail; its molecular structure and polymerisation are rather complicated and the description given here is much simplified. The cellular structure of plants is built up from cellulose, and the amount of **cellulose** varies from one type of plant to another; the long and fluffy-looking fibres in cotton plants are virtually all cellulose whereas most woods are only about 40% cellulose (the other major component being **lignin**, a network polymer that is undesirable in paper due to its more rapid degradation than cellulose). Cellulose is a polymer built up from glucose, a sugar, which has the chemical formula $C_6H_{12}O_6$ as shown in Figure 2.8(a) with one arrangement of the polymer in Figure 2.8(b). Again, the reactive –OH (hydroxyl) side-groups are involved in condensation reactions; you will notice that there are five of them in the glucose molecule. The glucose monomer can react to form a range of different polymers – starch, for instance, contains two polymers of glucose called amylopectin and amylose (shown in Figure 2.9). The first is a linear polymer; the second, which is the major component, has a branched structure. It is the many reactive sites on the ends of the branches that make starch such an effective adhesive. In cellulose the repeat unit is two glucose molecules which are bonded together with one unit turned upside down with respect to the other, as follows. This process is continued by further condensation reactions, each releasing a molecule of H_2O. A typical cellulose molecule contains as

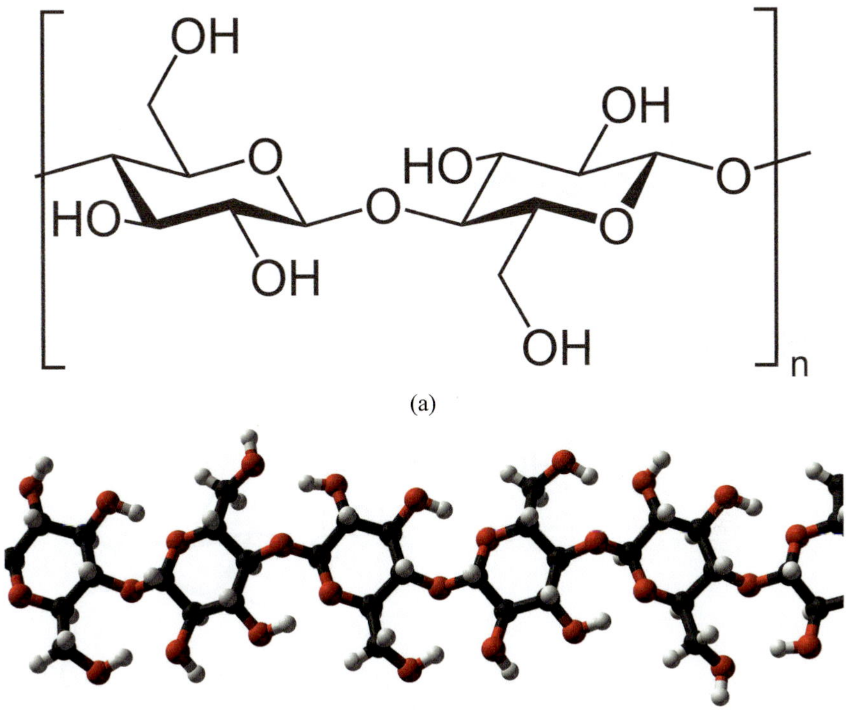

(a)

(b)

Figure 2.8 Structural formula for cellulose. Top: the monomer. Bottom: Schematic of the polymer.

many as 10,000 glucose units. It is the –OH (hydroxyl) on the left- and right-hand sides of the glucose molecule in the diagram that take part in the polymerisation reactions. In the cellulose polymer, each glucose ring still has three –OH groups. The hydrogen-bonding between these groups, which are highly polar, provides strong bonding between neighbouring polymer molecules.

The hydroxyl groups can be attacked by nitric acid (HNO_3) converting them to a nitrate group ($-NO_3$). Being bulkier than –OH groups, the nitrate groups force the polymer chains further apart. If all the hydroxyl groups are replaced in this way, the result is guncotton – an explosive. Controlling the nitration so that, on average, two of the –OH groups on each glucose ring are replaced by $-NO_3$ yields cellulose nitrate (CN), one of the earliest of the semi-synthetic plastics, first produced as Parkesine® and used as the base for early movie films. It is chemically reactive just like its explosive cousin, and even its suspected presence is concerning in social history and costume collections for its spontaneous flammability and its degradation product, which is nitric acid. The excess raw material remaining in factories after World War II was used to make nitrocellulose lacquer, which found a ready market in automotive paints applied by spraying, and this would later become **cellulose nitrate adhesive**, popular

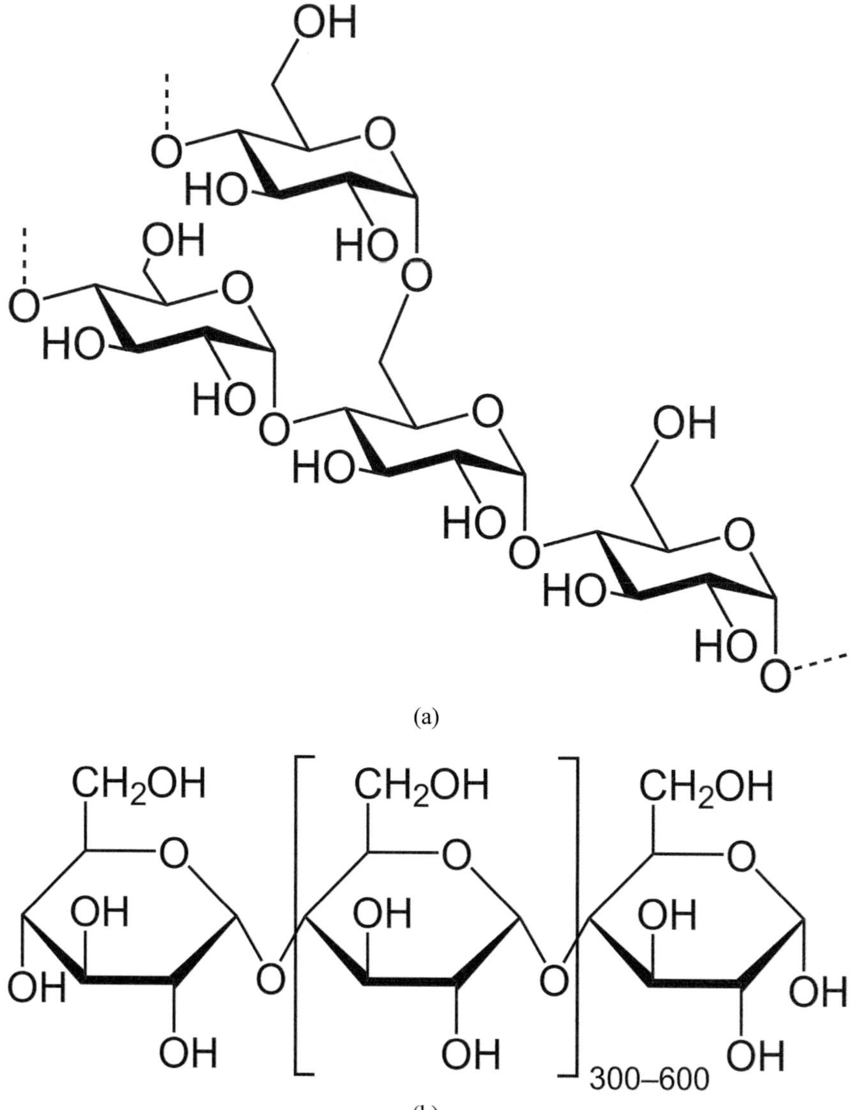

Figure 2.9 Structural formulae for amylopectin (left) and amylose (right).

at mid-twentieth century as a conservation material (popular amongst archaeological conservators to the extent that Paraloid® B-72 is today among conservators of most materials, but with less justification in terms of stability) and still perhaps to be found when old repairs are taken apart.

It should be emphasised again that using the structural formula of a polymer is adequate for the present purpose of depicting the covalent bonds in the chain, but that real polymers must actually 'look' very different, if we could visualise them directly at the molecular scale. They are three-dimensional, not two-dimensional; the chains are usually not straight; and depending on the size of the side-groups they can be jagged in appearance. Extensive studies of cellulose have shown that it is a long-chain molecule.

C Network polymers

In contrast to a long-chain polymer, a **network polymer** can form bulk material that is composed of a continuous three-dimensional molecule, the whole of which is bound together by primary bonds. Therefore, network polymeric materials have very different properties from polymers consisting of many long-chain molecules. The structure cannot be disrupted so readily by heat, solvents, or mechanical forces. There are many kinds of this type of material, both synthetic and natural, from semi-synthetic plastics such as Bakelite® found in domestic objects from the 1930s to the 1950s, to proteins, to dried oil paint. Such networks can be created in two basic ways:

- The linking of long-chains by atoms or small molecules.
- Interaction between small molecules to produce branched chains that join together.

An important example of the first method is the vulcanising of rubber. Natural rubber, poly-isoprene, is a hydrocarbon consisting of long-chain molecules (see Figure 2.10). These chains can be linked together by sulphur atoms that fit in between the long chains to join them at several points, and thus lock them together to some degree. Each sulphur atom forms a **cross-link** between two chains by forming a covalent bond with each; in this case using the double-bond available in the isoprene unit. The availability of a double-bond in the chain is necessary for this

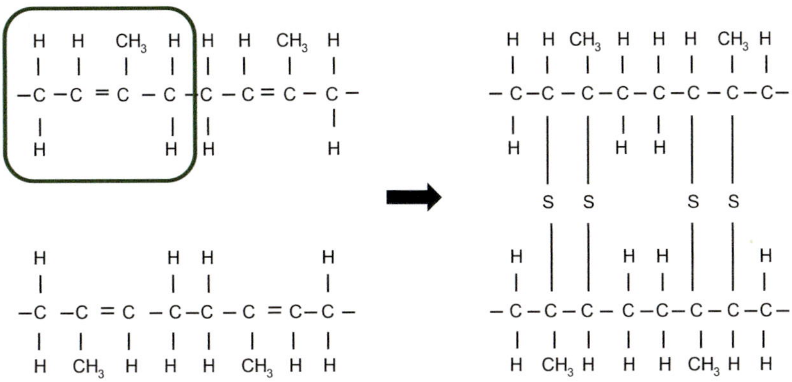

Figure 2.10 Schematic of one way of forming covalent bonds with cross-linking: polyisoprene (natural rubber).

method of building a network. Natural rubber with no sulphur atoms cross-linking it is soft and sticky, and prone to deterioration in the presence of the oxygen in air. A typical rubber used in vehicle tyres contains about 5% sulphur, which produces a network with about 500 carbon atoms between cross-links, and is damaged by oxygen far more slowly. A content of 40% sulphur produces many more cross-links and results in a hard, rigid, black, and glossy plastic used to make Vulcanite® and Ebonite®, and hence sometimes found in collections of musical instruments or earlier photographic technology. You can see from this that the degree of cross-linking has a dramatic effect on the properties of the polymer.

Similar long molecules can also be cross-linked to produce a network polyester. In order to do this the long-chain polyester molecules must be **unsaturated**, that is, contain double bonds. This means that the monomer from which the long chain is formed must contain a carbon-to-carbon double bond. One of the bonds in the double bond in the long chain can be used to create a cross-link with an identical site on a neighbouring chain. Just as sulphur atoms provide the cross-links in vulcanised rubber, a monomer such as styrene (Figure 2.11) can provide cross-links between the unsaturated long-chain polyester molecules. The result is a hard glassy material that is used as a casting resin, and also, in large quantities in glass-reinforced plastics (often know as GRP), used as a medium for some large-scale sculptures in the 1960s and 1970s. In practice, the long-chain polyester material is dissolved in styrene to form a viscous syrup. Just before use, the **hardener** (a catalyst), which initiates and maintains the cross-linking process, is mixed into it. The epoxy resins form network polymers in a similar

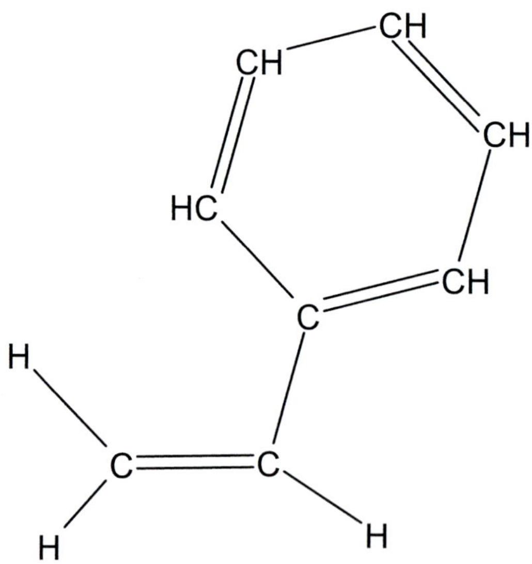

Figure 2.11 Structural formula for the styrene monomer.

way. In all of these examples, vulcanised rubber, polyesters, and epoxy resins, the cross-linking occurs by **addition polymerisation**, so no by-product is generated. Similar cross-linking of long-chain molecules also occurs in many proteins, for example in **collagen** (a main component of bone, tendon, skin, and hence leather) and in **keratin** (feather, hair, and horn).

The second way of producing network polymers, that is, by the interaction of small molecules that generate branched chains that link up, involves successive **condensation polymerisation** reactions. Examples are phenol formaldehyde (PF) and urea formaldehyde (UF) plastics, both of which are examples of earlier semi-synthetic plastics, used for industrial products and domestic objects at mid-twentieth century. UF has an interesting application in 'wet-strength' paper tissues and towels. The resin is incorporated in paper pulp during manufacture and forms a coating round each fibre. This prevents water saturating the hollow fibres but still allows absorption by capillary action.

The polymerisation of phenol formaldehyde can be considered as follows. Phenol, C_6H_5OH, has the structural formula shown in Figure 2.12(a), while formaldehyde, HCHO, is shown in Figure 2.12(b). These two monomers can react as shown in Figure 2.12(c). Then the same reaction takes place with some of the other hydrogen atoms surrounding the carbon ring, until a three-dimensional network is built up. As each link forms, water is again given off during the condensation reaction.

The basic types of polymer and the way they are formed has been established. You can now consider some of the characteristics that determine how they are used, particularly in conservation work.

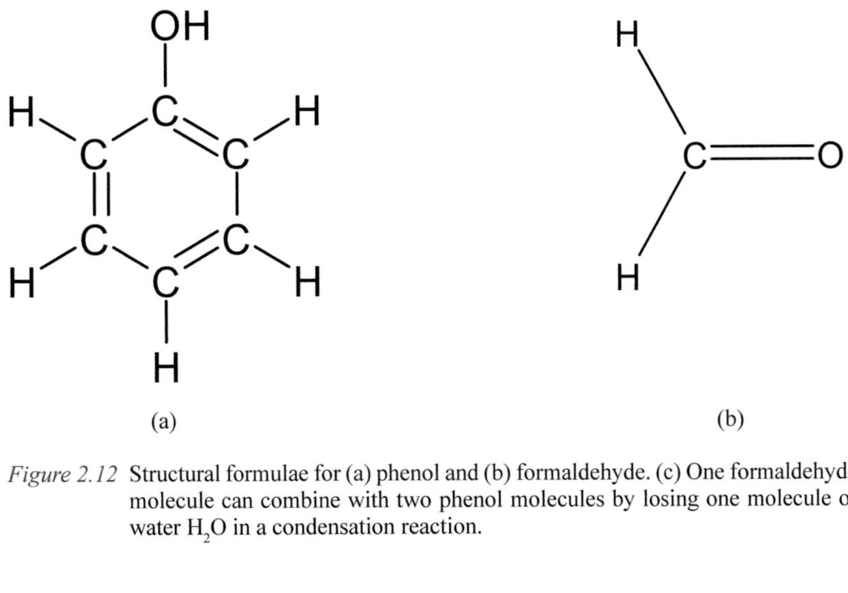

(a) (b)

Figure 2.12 Structural formulae for (a) phenol and (b) formaldehyde. (c) One formaldehyde molecule can combine with two phenol molecules by losing one molecule of water H_2O in a condensation reaction.

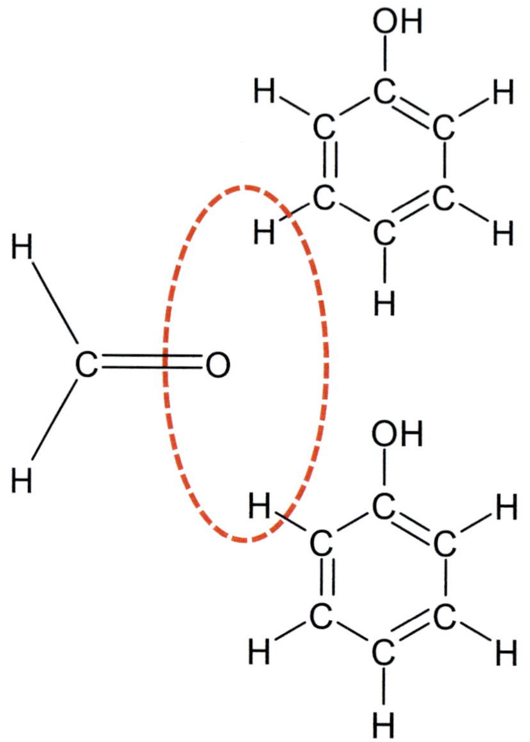

(c)

Figure 2.12 (Continued)

D Some important characteristics of polymeric materials

D1 Terminology

Of the vast numbers of chemical compounds that exist, a great many are polymers, and they range from the relatively simple synthetic materials like polythene and poly(vinyl acetate) and fairly straightforward natural materials such as cellulose, to the complex compounds that make up proteins, DNA, and beyond. All polymers can be considered in terms of the two types introduced in this chapter – long-chain and network polymers. However, a great deal of research and development on polymers has followed the commercial production of the first plastics products around the end of the nineteenth century, and this has led to the use of certain terms with which you need to be familiar – in particular **thermoplastics** and **thermosetting plastics** (or, simply, **thermosets**).

As its name implies, a thermoplastic is a material that becomes plastic, capable of being moulded, as it is heated. When these materials are warmed to a suitable temperature they become soft and can be formed into a desired shape. This is the basis of injection moulding. On cooling back to room temperature, they become rigid, but would soften again on reheating, repeatedly. Thermoplastics consist of

long-chain polymers, combined with various other molecules that modify their properties. A mass of long-chain polymer molecules is held together only by relatively weak secondary bonds between neighbouring molecules. Consequently, the molecules can be made to slide over one another relatively easily, particularly with increasing temperature. As you will see, the weak bonding between molecules influences many of the important properties of a thermoplastic.

A thermosetting plastic when fully polymerised, or cured, consists of a continuous network in which sliding between molecules cannot occur. In effect, the whole structure is one large molecule with all atoms linked by primary covalent bonds. The application of heat accelerates polymerisation and produces a permanent set in the material; hence the term thermoset.

In general, then, thermoplastics are based on long-chain polymers, and thermosets on network polymers. In conservation you will probably come across adhesives, coatings, and consolidants based on natural as well as synthetic polymers, and so you may find it useful to consider this statement the other way round: materials consisting of long-chain polymers have the characteristics of thermoplastics, while network polymers behave like thermosets.

Once a thermoset is fully cured it has set irreversibly. Such materials tend to be brittle and hard, and do not soften on heating; in fact, when heated they may char, burn, or crumble. How is it then that thermosets can be used as, say, resins for fibreglass, for moulded articles such as light fittings, and as adhesives, coatings, and consolidants? The answer is that when it is being worked, a thermoset is still a thermoplastic, that is, it consists of long-chain molecules that have not yet been joined together chemically. After the necessary shaping (moulding, brushing, spreading, and so on), the cross-linking process that produces the full polymer network is initiated. For thermosetting adhesives such as unsaturated polyesters and epoxies, this can be arranged very conveniently because the monomer that creates the cross-links can be packaged and stored separately until needed. Hence the two-component packs in which these materials are usually bought; one contains the pre-produced long-chain polymers, while the other contains the hardener (or **curing agent**) that produces the cross-links. Similarly, in traditional oil paints that contain natural fatty oils, such as linseed oil, atmospheric oxygen forms the cross-links between the oil molecules to form the dried paint film.

At this point, it is appropriate to re-emphasise the distinction drawn earlier between polymers and plastics. Polymers are very large molecules made up by the successive linking together of one or more small molecules (monomers); plastics are a synthetic blend based on one or more types of polymer, usually with additions of other materials designed to produce desired changes in properties – to improve their flow properties, resistance to oxidation, ease of handling, colour, and so on. Commercial adhesives, paints, lacquers, and consolidants are all examples of plastics. How some of these additions work will be discussed in the next section.

D2 Relating properties to structure

As a class, polymers provide a marvellous demonstration of how the properties of a material are determined by their internal structure. After all, combinations of atoms of only two elements, carbon and hydrogen, produce materials as different as methane, natural rubber, and polystyrene; add another element, oxygen, and the variety of products seems endless, from the simple molecule of formaldehyde to polymethylmethacrylate, polyesters, cellulose, fats, and many vitamins. This enormous range arises because the constituent atoms can be arranged in so many different ways. The internal structure of polymers can be conveniently classified at three different levels of scale:

- The types of atom and the way they are held together in the molecules.
- The size and shape of the molecule.
- How the molecules are arranged relative to one another.

Before exploring these levels a little, and their inter-relation, in terms of their properties, two terms need to be defined: **crystalline** and **amorphous**. When the atoms or molecules in a solid form a regular three-dimensional array, that solid is said to be crystalline. In such an arrangement the atoms or molecules are in their most stable positions, so crystallinity is the normal condition of solids. Common salt, sodium chloride (NaCl), is a very simple example.

Metals and alloys are aggregates of small crystals. Snowflakes are crystals, and so are diamonds, sapphires, and rubies. When allowed to form without physical constraint (for instance, a snowflake descending), the internal order of a crystalline solid is manifested externally by its symmetrical plane faces. These regular geometric shapes are what we normally think of as crystals. However, it is the regular internal structure that defines crystallinity and not the outward appearance.

On the other hand, a solid in which the arrangement of atoms or molecules has no regular order over any appreciable distance is called amorphous; the wide variety of glasses and glazes are examples. Some polymers are completely amorphous, some are mainly amorphous with a few crystalline regions, and others have a significant degree of crystallinity. In some forms of polythene, for instance, more than half of the volume is crystalline. The degree of crystallinity in a polymer is an important factor in determining its properties, so it is useful to look in a bit more detail at why crystallinity can vary so much. Consider, for instance, what happens to a thermoplastic (a mass of long-chain molecules) as it is cooled from the molten (liquid) state.

As you will remember from *An Introduction to Materials and Chemistry*, the liquid state is characterised by an arrangement of molecules which is highly disordered and therefore relatively less closely packed. As a consequence a liquid has a lower density than the equivalent solid made of the same material (but now at a lower temperature), and also a lower viscosity, that is, the molecules in the liquid occupy more space and, because they are less strongly bound, slide over one another more easily. (Water and ice are a notable exception here – for a given volume of water, the ice has a greater volume, because hydrogen bonding dominates in this molecule.)

As the thermoplastic cools below its melting point, at which temperature the solid form becomes more stable than the liquid form, the molecules attempt to attain the ordered arrangement of a crystal. The atoms or molecules in a material are in a constant state of motion due to the thermal (kinetic) energy they possess, and

Table 2.1 Some examples of crystalline and amorphous polymers.

Crystalline	*Amorphous*
polyethylene (PE)	polymethylmethacrylate (PMMA) and its co-polymers
saturated polyesters	PVAc-PE co-polymers
cellulose	polystyrene (PS)
cellulose acetate (CA)	all thermosets
polyamides (for example, nylons)	

the higher the temperature the greater is the thermal energy. The regular arrangement of a crystal cannot occur instantaneously on cooling because it takes time for the atoms or molecules to move to an appropriate place in the ordered arrangement. In practice, crystalline regions begin to grow at various points or centres of reaction (called nuclei (which means, a centre), but not to be confused with atomic nuclei) in the liquid, and they grow into the liquid as further atoms or molecules move to join the regular array. The final result of this process would be a solid consisting of a mass of small crystals; a **polycrystalline aggregate**, as it is called. This is what happens in metals because the single atoms involved are sufficiently mobile to fit into a regular array. In polymers, however, the same configuration is rarely, if ever, attained. There are two reasons for this. Firstly, long molecules have a very limited mobility in the liquid compared with single atoms. Secondly, even if they could move freely, the size and shape of the molecules often militates against *all* of the molecules lining up like sticks in a bundle. If bulky side-groups are attached at irregular points along the molecule, they are unlikely to pack together easily. For example, random co-polymers are less likely to crystallise than block co-polymers. On the other hand, strong secondary bonding between polymer molecules encourages the development of crystallinity. Table 2.1 lists some of the common polymers that are either amorphous or significantly crystalline.

Notice that the lists bear out the general points made earlier about molecule shape. Those that give rise to strong secondary forces between molecules due to polar side-groups, especially C=O, often take part in hydrogen-bonding (to –OH groups, for instance) and tend to have a degree of crystallinity; those with small regularly spaced side-groups behave similarly, but those with large side-groups are usually amorphous. Where there is an irregular structure but no polar side-groups the polymers tend to be 'rubbery' (for example, natural rubber). To help you visualise the molecular arrangements in long-chain polymers, Figure 2.13 illustrates the type of model that has been used to describe an amorphous and a highly crystalline material.

Drawing a polymer out into a fibre increases its crystallinity – just as stretching a piece of cotton-wool aligns many of the fibres in the direction of pull. Injection moulding or extruding a polymer to shape it also induces some crystallinity. As Figure 2.13(b) illustrates, even a highly crystalline polymer contains some amorphous regions.

From the prior discussion of the problem of crystallisation it is clear that amorphous polymers have such awkward molecular structures and high viscosity that

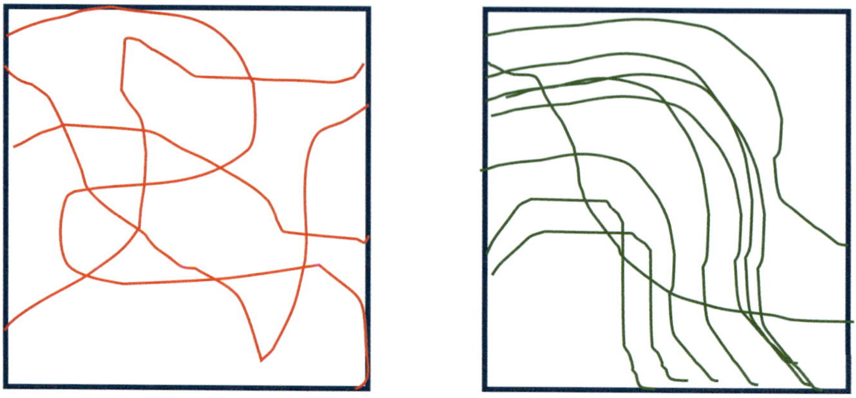

Figure 2.13 Schematic showing (a) the tangled chains of an amorphous polymer and (b) the more ordered chains of a crystalline polymer.

they do not crystallise at all. When cooled, they exhibit the properties of a solid, such as becoming more rigid and dense. Unlike crystalline materials, they do not show a sharp melting point as they change from solid to liquid on heating. In this way they are like glasses which alter gradually from being like a solid material to being like a liquid over a range of temperature. The lower limit of this temperature range is called the **glass transition temperature**, and the conventional symbol for it is T_g. Essentially, at this temperature the available thermal energy is small compared with the forces holding the molecules together. At lower temperatures, very little molecular adjustment is possible, in other words, the molecular structure is 'frozen in'. For example, below its glass transition temperature by more than 10°C or so, an amorphous polymer is typically hard and brittle; above T_g by about the same amount it is typically softer and more rubber-like, is more readily permeated by gases and can be dissolved more easily.

Clearly, the glass transition temperature of a polymer is extremely relevant to the uses of the polymer. The values of T_g for many of the polymers referred to in this text are given in Table 2.2. Notice that thermoplastics that are classed as crystalline also have a T_g because they invariably contain some amorphous regions.

You can see that in their raw state, materials such as cellulose acetate, PMMA, and PVC are hard, brittle, and glasslike at ordinary temperatures. However, long-chain polymers like these can be mixed with other chemicals which effectively decrease T_g; that is, they induce rubber-like properties, such as enhanced flow, at lower temperatures.

Such substances are called **plasticisers** and are relatively involatile solvents for the polymer. By being incorporated between the molecules of individual polymers and weakening the *inter*-molecular bonds, they separate the long chains and facilitate their movement relative to one another. Plasticisers need to behave like a solvent in order to carry out their function and they also need to have a low vapour pressure – so that they do not evaporate, or do so only slowly – which is fine for the

Table 2.2 Approximate glass transition temperatures for some common polymers (rounded to the nearest 5°C).

Polymer	T_g *(°C)*
Polyethylene (PE)	−35 to −90
Natural rubber	−75
Paraloid® B-72	40
Nylon 6	50
Regalrez®1126	65
Polyvinyl chloride (PVC)	85
Polystyrene (PS)	95
Polymethylmethacrylate (PMMA)	105
Cellulose acetate (CA)	105

lifetime of a commercial product, but which can become important over its long afterlife in a museum collection. (They normally have a molecular mass greater than 300.) Clearly, the evaporation of a plasticiser leads to brittleness – an ageing problem often met a few decades after application of adhesives, coatings, and consolidants. A few practical examples should help to emphasise the importance of a plasticiser. Cellulose nitrate is a hard brittle thermoplastic at ordinary temperatures; its T_g is similar to that of cellulose acetate, about 100°C. Mixing it with camphor as the plasticiser led to developments of the first semi-synthetic plastic, namely Parkesine® and, later, Celluloid® based on the same polymer.[2]

PVC in its pure form is one of the hard, rigid plastics you come across in the clear plastic bottles used to contain bottled water and soft drinks. When plasticised, it becomes soft and pliable and is used in 'leathercloth' for book bindings, car seats, 'pleather' fabrics used in clothing as a vegan alternative to leather, and so on. Common plasticisers for PVC are esters formed between ortho-phthalic acid and alcohols containing about eight carbon atoms.

Manipulating the formulation of a plastic can reach a high degree of subtlety. For instance, as you will see in Chapter 6, the T_g of an emulsion paint should lie, initially, some degrees below internal temperatures when the paint is being applied, so that a continuous paint film is obtained. Ultimately, however, the film may need to be hard and glasslike, that is, to have a T_g above ordinary room temperature. This is achieved by incorporating a small amount of fugitive (volatile) plasticiser in the thermoplastic polymer paint base. During the final stages of drying, this plasticiser is slowly lost by evaporation, thus producing a rise in T_g.

Finally, in this brief review of the structure/property relationship in long-chain polymers, it is important to appreciate that the degree of crystallinity has a marked effect on properties. This is basically because in a crystalline region the molecules are packed very closely together compared with the arrangement in an amorphous region, which means that the spaces between molecules are smaller and the binding forces stronger in a more crystalline region. As a consequence, in general, a crystalline polymer is expected to have, for example, higher strength and rigidity than the equivalent amorphous polymer and to be less easily dissolved and less readily permeated by gases. Over time, the degree of crystallinity in a polymer declines, and measurement of its average value has been used as an indicator of degradation in polymers in museum collections, especially cellulose acetate.

Of course, network polymers (thermosets) are very different to long-chain polymers (thermoplastics). By their very nature, they are amorphous, and since they consist of one continuous molecule, they do not pass through the glass transition range in the same way as thermoplastics. Their mechanical strength, rigidity, and hardness are determined essentially by the spacing between the cross-links along the molecule: the more frequent the cross-links, the more rigid the material, and the more insoluble it becomes.

D3 Optical properties

Many polymers, provided they are free of manufacturing residues and also not severely aged, are naturally colourless and transparent. This is why Perspex®/Plexiglas® made from PMMA is used for display cases, sometimes for glazing paintings or watercolours, and as the transparent sheet for face-mounted photographs. This particular polymer is used because its ageing properties are useful for this purpose; it does not yellow or lose its **water-white clarity** over many decades. The term 'water-white clarity' means that a block of polymer even several millimetres in thickness does not give a colour cast when it is placed over a colourful image. Departure from this desirable property can be monitored by **colour measurement** repeated over a longish period, up to the point when the yellowing becomes noticeable to conservators, who are accustomed to looking for small colour changes. Water-white clarity is also a desirable property in an adhesive used to join together naturally transparent or translucent materials such as glass or Perspex®/Plexiglas®, and it provides a substrate with no colour compromises to make retouching a challenge, when losses in a glass vessel have to be cast in and then toned to match the glass body. Acrylic adhesives such as Paraloid® B-72 also have water-white clarity and in addition are stable when exposed to light, and these two properties account for the popularity of this material as both adhesive and coating. (Note that a coating or a block of cast resin may be coloured yet transparent, while an aged and yellowed one may still look transparent, so the description 'transparent' in itself is not sufficient to define desirable appearance in an adhesive, coating, or consolidant.)

The ideal adhesive for butt-joining two transparent fragments should look as if it isn't there at all, and indeed if there had never been a fracture requiring repair. Precisely matching the **refractive index** (**RI**, written as 'n' in equations) of the adhesive and the transparent materials being joined would achieve this ideal. Refractive index, described in more detail in *Colour, Colour Measurement and Colour Change* in this series, is a measure of the way a ray of light is bent from its original path when it goes through a transparent material. The RI of air is 1 by definition, because we live and breathe air and do all our seeing through unnoticeable, water-white blocks of air without even thinking about this. The range of possible values of RI sounds pretty small: animal glue in water has an RI around 1.33; PMMA and acrylic adhesives have an RI of 1.48; materials like chalk or powdered silica that might be used to matt down a transparent coating have an RI of about 1.5; the medium of oil paint has an RI of 1.50 rising to 1.52 or more as it ages; natural resins are about 1.55; leaded glass used to make decorative glass might have an RI approaching 1.6 and specialist scientific glass components perhaps 1.7 or 1.8; the very opaque pigments lead white and titanium dioxide have RIs of 1.93 and 2.00 respectively; diamond is about 2.4 and exceeds the RI of all other gemstones. In this context, Paraloid® B-72 at 1.48 and leaded glass at 1.60 are pretty mismatched in RI, whereas other potential adhesives such as epoxy resins or polyester resins are a better match to such glass, their RIs being typically 1.55–1.56, among the highest that can be found among conservation materials past and present.

In terms of selecting an adhesive to join broken fragments of glass, choosing the right adhesive properties is important as will be discussed later in this book, but optically the hierarchy of importance is likely to be: water-white clarity > matched adhesive properties > refractive index.

The opposite description to transparency is opacity, and in fact opacity is simply a way of stating that the opaque material appears so because the medium it is in (usually air) has a widely different RI. Powdered chalk before it is added to matt down a coating is one example, and chalk functions as a fairly opaque pigment when it is used in watercolour paint, but as fairly unnoticeable paint extender when it is added to oil paint, and so immersed in a medium of higher RI. Glass fibre-reinforced polyester resin (GRP) also illustrates the concept: glass fibres sitting in a jar or on the bench look like fine white threads: when added to polyester resin they do not react chemically but their presence makes it a stronger composite material than the resin alone, and their close RI to the resin means that they are not noticeable at all. (This is not advocating this composite material as a conservation adhesive! However, it has been used by some modern sculptors and is found in engineering, model- and scenery-making, and construction contexts, so it can be found in some mixed museum collections and may require conservation treatment.) It begins life transparent and fairly colourless, but yellows on ageing and eventually acquires a green/brown and slightly muddy or **translucent** appearance; that it, it becomes more opaque looking. Translucency is a very subjective term that can be used to describe anything from the slightly cloudy greenish appearance of excavated and cleaned Roman glass in exceptionally good condition to opalised Perspex® which is pretty close to being opaque even if held in front of a bright light. It's not a common or very helpful term in a conservation context for this reason. Some materials such as GRP undergo changes in appearance that are quite challenging to describe in words, and which may be conveyed more clearly by colour measurement, or **gloss measurement**. Like translucency, gloss can be a term clearly defined only in the mind (or eye) of the beholder: it is prized on much traditional wood furniture, and restoring gloss may be an aim of a consolidation or coating treatment, whereas introducing a glossy appearance during the consolidation of matt and underbound paint would represent a failed treatment and a damaged object. Thus 'glossy' may sometimes be a fairly unhelpful term too, although gloss measurements or a treatment practised on mock-up material to assist in the best choice of coating or consolidant can be very helpful.

Notes

1 Van Oosten (2022) gives a very comprehensive list of the polymers conservators are likely to encounter in plastic-based objects, with date of first manufacture, proprietary names for these plastics, applications, appearance and likely degradation, and physical properties.

2 Kim *et al.* (2024) illustrate a variety of degradation problems for cellulose acetate objects and discuss the selection of adhesives and consolidants for their treatment.

3 Making joints with different types of adhesive

Now that some of the chemistry of polymers has been explained, it is possible to look in more detail at the range of adhesives which are available for joint-making. Since this book is not an adhesives manual, by no means will all individual adhesives be covered, but the three classes of adhesives – which embrace all the materials used in conservation – will be discussed. They are solvent, reaction, and melt-freeze adhesives.

A Solvent adhesives

> No one should use a solvent-based adhesive and nobody should make one. They contain only one-quarter to one-third of useful adhesive; they are inflammable, toxic, foul-smelling and expensive since you are going to throw over half the bulk away by evaporation. Solvent-adhesives are legislated against, one by one solvents are banned.
>
> (Harrison 1979)

In spite of these very appropriate thoughts, **solvent-based adhesives** have been and still are widely used in conservation and therefore warrant serious consideration in this text. As coatings, they remain in common use too. Concerns about greenness and sustainability have prompted research into less harmful solvents that will still permit the application of useful polymers as adhesives and coatings, as an alternative to those based on petrochemicals.

The principle on which solvent adhesives are based is simple: the adhesive is dissolved in a suitable solvent so that it can be applied to a surface in liquid form. The solvent then evaporates to leave the solid joint. The adhesive bonds so formed are mainly due to secondary forces between molecules in the adhesive and those of the surface. The considerable advantages of solvent adhesives in conservation work are that unlike melt-freeze adhesives they can be made to spread very easily (because of their very low viscosity), and, unlike reaction adhesives, the joints are more readily reversed, both during treatment and when re-treatment in the future becomes desirable for another reason. On the other hand, in addition to the disadvantages cited in the earlier quotation, perhaps their greatest shortcoming is the

DOI: 10.4324/9781003389217-3

shrinkage and embrittlement that can occur during ageing owing primarily to the continued loss of solvent, a few percent having been retained by the polymer after apparent setting. Examples of solvent adhesives commonly used in the twentieth century and still today are cellulose nitrate, poly(vinyl acetate) and its co-polymers, hydrolysed starch and animal glues (various gelatines produced by the breakdown of collagen, a component of bone, cartilage, and skin).

A1 Cellulose nitrate

Cellulose nitrate was one of the first polymers to be widely made and used. It belongs to the class of semi-synthetic polymers, and it was the first plastic base for movie film. Many early plastic objects and utensils were moulded using Celluloid®, the commercial name which is most commonly associated with cellulose nitrate. In conservation its use stretches back over many decades, and because of tradition and the acceptable results achieved, it may still be used as an adhesive for archaeological pottery. As a coating, it has long been used on silver in an attempt to reduce the rate of tarnishing in storage or on display.

The polymerisation of glucose monomers to form long-chain molecules of cellulose was described earlier, and it was shown that cellulose nitrate is formed by the reaction of nitric acid with some of the hydroxyl (–OH) side-groups of the cellulose molecule. Part of a cellulose molecule is illustrated in Figure 2.8; cellulose nitrate is similar but with about two-thirds of the hydroxyl side-groups replaced by nitrate ($–NO_3$) groups.

The solvents most commonly used in cellulose nitrate adhesives were mixtures of acetone, ethanol, and butyl acetate. After many years, it is still found to be soluble, and therefore the joints are reversible. There is something to be said for older, well-tried, and tested materials: by now we know from observation in real time the best and worse they are capable of. However, cellulose nitrate adhesive has three major drawbacks: brittleness and shrinkage, as well as the evolution of nitric acid when it breaks down.

The hydrogen bonding between the –OH groups on neighbouring cellulose molecules is strong, and this is reduced in cellulose nitrate by the presence of the $–NO_3$ groups. Nevertheless, solid cellulose nitrate is a relatively rigid material and is brittle. It was usually used without any additives, but it can be made more flexible by the addition of a plasticiser. Camphor was the plasticiser historically used for cellulose as a plastic. This helped to solve one problem but caused others. With age, the camphor slowly volatilises, which means that the joint becomes more brittle and also shrinks. In certain storage situations, with large collections of objects thus treated or made from cellulose nitrate enclosed in rolling racks for example, there is a theoretical health hazard to staff accessing the enclosed space after the camphor has been evaporating for long periods.

In common with other solvent adhesives, cellulose nitrate shrinks during setting because of the loss of solvent. The result of this is that joints made with cellulose nitrate are rather weak and brittle, but in many conservation applications, such as the repair of earthenware, they are quite satisfactory.

A2 Polyvinyl acetate

PVAc must have been one of the most widely used adhesives in the second half of the twentieth century, and it has also found conservation applications in consolidation and coating. As an adhesive it has been used for support fabrics in textile conservation, for mending ceramics and in stone conservation for example. Like several classes of polymer (**acrylics**, polymers of acrylic acid: $CH_2 = CH (COOH)$, are another common example), PVAc can be used as a solution or an emulsion. Its use as an adhesive will be discussed in these terms. Sometimes the properties of an organic solution will be preferred; at other times an aqueous emulsion will be more suitable for the job in hand.

As noted earlier, calling this material PVA for short, as is often done for commercial products, can be misleading because the abbreviation is also used to describe poly(vinyl alcohol). It is better to call them PVAc and PVAl respectively. In its pure form PVAc forms a clear glassy solid with a high resistance to heat, light, dilute acids, dilute alkalis, and water. It dissolves readily in many organic liquids, and acetone and ethanol are commonly used as solvents for the adhesive in solution. Because PVAc is a thermoplastic of good chemical stability, joints made with it are easily reversed by dissolution. Like cellulose nitrate, it is fairly brittle, but a small amount of solvent is retained for a considerable time and acts like a plasticiser. (Blocks of PVAc dissolved in ethanol and left to harden have remained flexible for several years.)

A number of PVAc adhesives have long been available commercially (usually labelled 'PVA adhesive') and recommended for use on wood, paper, and fabrics. These materials are usually in the form of an **emulsion** or **dispersion**. An emulsion consists of finely dispersed particles of the adhesive polymer in suspension in water. The best known example of an emulsion is milk, which consists of minute particles of fat in suspension in an aqueous medium.

A3 Acrylic adhesives

By far the best-known acrylic adhesive today is Paraloid® B-72, a co-polymer of ethyl methacrylate and methyl acrylate (EMMA). It has the best ageing properties of the acrylic adhesives since it does not yellow or change in solubility over decades in indoor museum conditions, does not react with any materials being conserved unless they are sensitive to the solvent used to dissolve it for application, does not require a plasticiser, and is soluble in a wide range of solvents including ethanol, acetone, xylene, and toluene. (This is in written order of harmfulness, ethanol being the least harmful solvent and the only one likely to be produced in future from green ingredients. The others are made from petrochemicals.) It is thus a good candidate for investigations into alternative and greener solvents, since the list implies quite a range of solubility parameters for the polymer. This offers many possibilities for application by brush (as adhesive, coating, or consolidant) or spray (as a varnish, in a suitable spray booth with fume extraction), with some control of drying time. Many conservation suppliers have sold and still sell ready-dissolved

Paraloid® B-72 in a tube, which is convenient for work in the field, where materials all have to be taken to site ready for use. Today this is usually labelled by its own name and as 'conservation adhesive'. In the past, a popular UK product was sold as HMG® glue. Incralac® which incorporates a corrosion inhibitor today regarded as harmful to human health is still available, marketed as a combined coating and corrosion inhibitor for copper, brass, and other copper-containing alloys.

In the later decades of the twentieth century, Paraloid® B-67, based on isobutyl methacrylate (iBMA), was almost equally popular as a coating and consolidant. It was however found to yellow over time and to become re-soluble in a narrower range of solvents over time, due to cross-linking. Its use today would be unusual as a coating or varnish, but it must survive on many objects, and especially those that were last treated decades ago in private studios rather than in public museums, where its use (in Europe at least) persisted for longer.

At the same period, a conservation coating based on Paraloids® B-44 and F-10 was popular for coating and protecting objects, including metals. Its tendency to yellow was moderate, but today Paraloid® B-72 is a more widely used product.

Demonstration

Pour some cooking oil and some water into a bottle and shake vigorously. You will notice that the oil breaks up into small droplets that are dispersed in the water. However, the droplets soon coalesce and the water and oil separate as two distinct layers; the same happens when cream floats to the top of a bottle of milk. Now add a small amount of washing-up liquid (or a soap solution) and you will find that the emulsion lasts for much longer. As you know from *Cleaning*, soap molecules are surfactants; they have a hydrophobic (water-repelling) end and a hydrophilic (water-attracting) end. The soap molecules diffuse and collect on the oil droplets and act as a barrier that inhibits coalescence of the droplets. This is illustrated in Figure 3.1 in which soap molecules are represented by a convention which you will remember from *Cleaning*.

Adhesive emulsions behave in a similar way to the mixture of oil, water, and soap solution. As the water evaporates from the emulsion, the particles of polymer coalesce to form a continuous solid. To form an homogeneous solid the polymer molecules must be able to move and adjust their positions. This means that the polymer must be a thermoplastic and above its glass transition temperature, T_g. Of course, to make a good joint, the resultant solid needs to be rigid, and this requires the polymer to be below its T_g. As pointed out earlier in Chapter 2, this change is achieved by incorporating a small amount of plasticiser that evaporates as the adhesive dries out.

In the case of PVAc emulsion adhesives, the surfactant usually used to **stabilise** the emulsion (stop the polymer from settling out) is poly(vinyl alcohol) (PVAl), which with its –OH groups is hydrophilic. Some plasticiser is also added. The solids formed from aqueous emulsions are inevitably sensitive to moisture, due to the presence of the hydrophilic stabiliser, and this necessitates the addition of fungicide and bactericides – and restricts the use of PVAc emulsion adhesives to applications indoors.

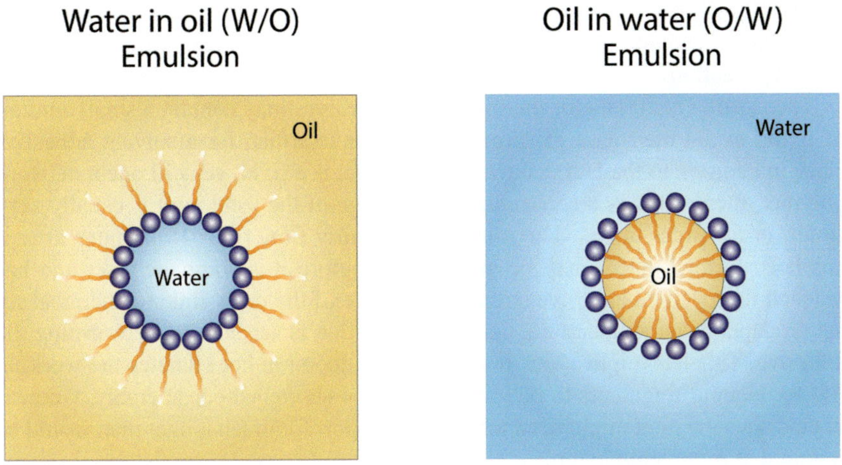

Figure 3.1 Water-in-oil emulsion (left) and oil-in-water emulsion (right). This structure is called a micelle.

Image: Ph-HY via Shutterstock.

Emulsions such as these are very useful because they allow much more control over the polymerisation process and, therefore, the properties of the adhesive. But, as usual, there are swings and roundabouts. The addition of these chemicals increases the susceptibility to oxidation and possibly the formation of cross-links and increasing insolubility of the adhesive. The *volatilisation* (vapourisation) of these chemicals, especially in the confined atmosphere of a display case, may also lead to corrosion of metal objects on display.

B Reaction adhesives

There are many instances where the use of solvent adhesives is unsuitable. This occurs most often because they set too slowly, shrink too much and do not give joints of sufficient strength. Although strength is not, for its own sake, important in conservation, it is necessary that the weight of a heavy ceramic or metal object will be supported by the joint. Dowelling with stainless steel rods may accomplish this in some cases but it is also often necessary to use **reaction adhesives**.

These adhesives are produced *in situ* by chemical reactions and adhesion is due mainly to strong secondary bonds. They are usually thermosetting (network) polymers and are formed by one of two main methods: one in which the cross-linking monomer and other chemicals to initiate and control the network formation are kept separate from long-chain polymers and mixed when required; and one in which polymerisation is inhibited until the monomer makes contact with the surfaces to be joined. Epoxy resins are examples of the former, and alkyl cyanoacrylates (often sold as **Superglue®** or some variant on such a name, to imply almost superhuman strength) are examples of the latter.

Not all reaction adhesives are made from organic polymers. For instance, plaster of Paris, cement, and mortar also solidify *in situ* by means of chemical and physical reactions.

The starting materials for thermosetting adhesives may contain a small amount of liquid to aid their ease of spreading, but far less than for a solvent adhesive. This, in addition to the fact that the solid product is a rigid, yet still open, network polymer, means that as a consequence shrinkage of the solid joint is usually very small, in the order of 1%. The joint may be fairly firm 10 or 15 minutes after it has been formed, but it may be many hours before it is completely set and has reached its specified adhesive strength. The rate of the cross-linking chemical reaction depends on temperature, and so the reaction is accelerated by warming the adhesive. This has implications if it is used outdoors: in hot climates the working life for placing fragments to be joined would be shortened considerably, whereas in cold sites the joint might need temporary support for much longer than would be required at room temperature.

However, most of the polymerisation reactions in this type of adhesive are **exothermic** – they give out heat; the heat released accelerates the process. In selecting such an adhesive, consider whether the heat evolved could damage the object – by, for instance, causing evaporation of moisture and thus causing internal stresses.

B1 Epoxy adhesives

An epoxy resin adhesive (or epoxide adhesive) is made in two stages. The first step is to produce relatively long-chain molecules by condensation polymerisation reactions. Typical monomers in these reactions are shown in Figure 3.2. They are often known by confusing non-systematic names, such as epichlorhydrin (Figure 3.2(a)) and Bis-phenol A (Figure 3.2(b)). Bis-phenol A is found in most of the relatively or completely water-white epoxy resins that have found application in conservation.

The initial reaction is, in effect, a condensation reaction between these two molecules followed by further additions in which the three-membered ring in epichlorhydrin is opened to form an open chain with a hydroxyl –OH group on it. The product of this reaction is the so-called resin component (or part A) of a two-pack epoxy resin system. In the second step – the one that is carried out *in situ* by mixing the resin with the hardener (part B) and immediately deploying the resin as an adhesive – these linear chains are cross-linked by addition polymerisation reactions using another monomer. The monomer in the cross-linking stage is usually a di- or polyamine, a compound containing two or more $-NH_2$ groups, such as diethylene triamine ($NH_2CH_2CH_2NHCH_2CH_2NH_2$). The cross-link is produced when the amine group reacts with two epoxy groups. The COC three-membered ring is called the epoxy group and compounds containing it are called **epoxides**. Figure 3.2(a) shows the reaction of one end of a polyamine with the two epoxy end groups of the long-chain molecules. Similar reactions will take place at the other amine groups on the polyamine, and this causes the build-up of a network.

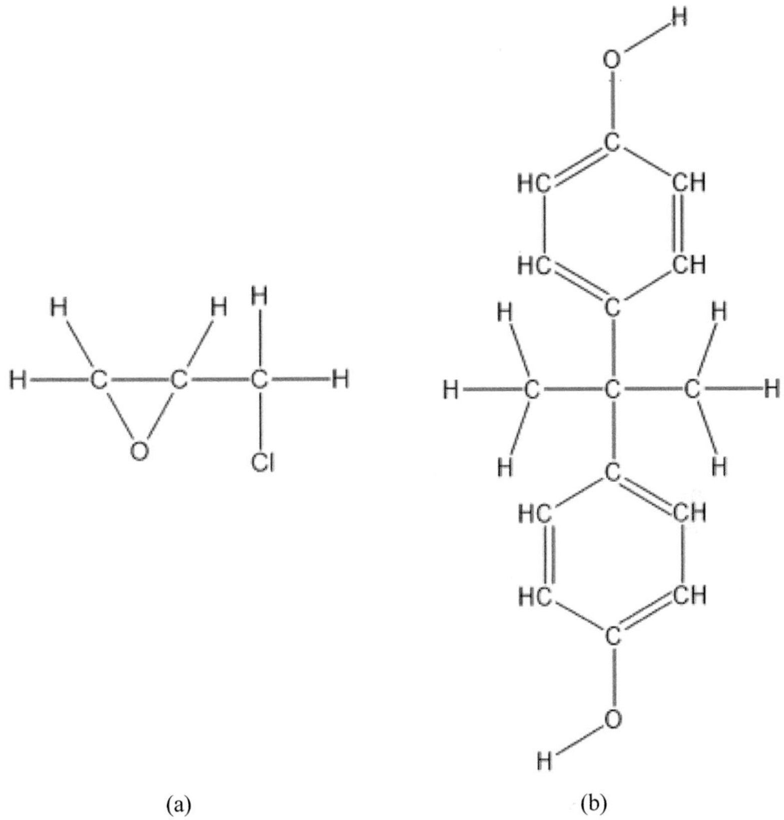

Figure 3.2 Structural formulae for (a) epichlorhydrin and (b) Bis-phenol-A which can be reacted to form epoxy resin.

Many different resins and hardeners can be used and the number of permutations of these is very large. A manufacturer often sells a wide range of different epoxy systems to cover all sorts of operating conditions; different temperatures, different setting times, and different viscosity. Tailoring and mixing by the user is possible but should only be attempted by a conservator confident about the chemistry involved. The systems with the fastest setting times may retain air bubbles from mixing, whereas a resin with longer **working time** (also known as 'pot life') before it begins to gel usually releases any bubbles. These would obviously cause points of weakness in a join if they remained in place.

Rather few epoxy adhesives are completely colourless, and this may be a disadvantage, particularly when joining transparent materials. The truly water-white ones contain fewer manufacturing residues, and in consequence have better resistance to yellowing and (also in consequence) much higher price tags. Because of the nature of the production processes, colour and other properties may vary between batches, so it is important not to rely too heavily on the characteristics of one sample – and not to use up a poor batch of resin for the sake of economy. Some of the starting materials deteriorate with time in their containers because the ingredients undergo chemical changes: there is a definite storage lifetime for two-part resins. These reactions can be inhibited by keeping the containers in a refrigerator – the lower the temperature, the slower is a chemical reaction (see *An Introduction to Materials and Chemistry*, Chapter 2).

To achieve maximum rigidity and strength of the adhesive, it is essential for the resin and hardener to be mixed in the recommended proportions. Weighing the resin in its mixing pot and then adding the correct weight of hardener by using a dropper or disposable pipette is much more accurate than measuring the two parts separately by volume and then attempting to pour the entire contents from a small measuring cylinder into the mixing pot.

Room-temperature curing does not immediately achieve the full bonding potential of the adhesive. For the polymeric network of the thermoset to be developed, that is, for cross-links to be formed, the cross-linking agent has to be mobile and diffused through the resin. Thus the molecules of the resin must be able to move to accommodate the cross-linking monomer – it must be above its T_g. As more and more cross-links are formed, the material becomes more rigid and its T_g is effectively raised. Diffusion is therefore inhibited and the cross-linking reaction stops. Raising the temperature above T_g permits the reaction to resume. In industrial applications, curing is often carried out in two stages; the first at room temperature which produces partial curing sufficient to hold the pieces in place, and the second in a higher temperature environment that produces the required degree of cross-linking and thereby strength and stiffness. The only ethical conservation equivalent is to independently support the joint at room temperature for longer – perhaps five times longer – than the stated setting time for the product used.

Bis-phenol A type epoxies have been used since the later twentieth century to support joints that have to be load-bearing (such as those in furniture, ceiling paintings, and frames for large and heavy stretched canvases) or to bridge gaps and replace missing pieces as well as join fragments of glass together. (Different types of epoxies find industrial uses in challenging environments like the repair of buried and undersea piping, but their dark colour and rather aggressive chemistry preclude such products from conservation use.) Fully developed joints formed with epoxy adhesives are very strong. Metal has a greater breaking strength than epoxy, so the adhesive would fail first, but when epoxy is used to join wood, and aged and degraded wood in particular, it is the wood that would fail first: this is not a good outcome from a treatment. Epoxy resins are also insoluble in common solvents, so such joints are difficult to reverse. Dichloromethane may, as a result of slow diffusion, eventually produce sufficient swelling to disrupt the solid resin, but this is a harmful solvent to be used only with suitable **personal protective equipment (PPE)**. This usually means a solvent vapour mask if the operation cannot be carried out in a fume hood, gloves that are appropriate for the solvent in question and unaffected by it, and sometimes eye protection (which may be provided by correct used of the fume hood). Soaking in water for a few days will remove epoxy resin from the mock-ups it was trialled on. Neither of these scenarios could be applied to many objects, while commercial epoxy 'digesters' sold for the taking down of epoxied joints would be extremely damaging: epoxy resins are not reversible in practice and should only be selected in the knowledge that they will reduce future treatment options.

Epoxy resins have good **cohesive strength** as well as adhesive strength and provide a better match to the mechanical properties of inorganic materials such as glass, metals, and stone than any solvent adhesive. The high cohesive strength means that they can be used to cast missing fragments and to bridge gaps in a broken object. Well-preserved or pristine glass or metal joined by epoxy adhesives offer a better possibility for reversal of a past treatment.

Users of epoxy resins should also be aware that epoxies act as allergic sensitisers for some individuals, so risk assessments in open-plan or shared studios need to take account of this. **Allergic sensitisers** affect an individual by producing troublesome or even dangerous reactions such as itching, eczema, and breathing issues following second and subsequent exposure to the offending material, even to tiny amounts. That means in practice that someone in the same workspace, or working near the fume extraction outlet, could be seriously affected by someone else using the material correctly, protected by PPE. It is better to use more PPE from the first use of such materials, to limit the possibility of being sensitised, and to be aware that others could suffer a strong reaction to the material.

B2 Polyester resins

Polyester resins are synthetic resins formed by the reaction of dibasic organic acids and polyhydric alcohols. Maleic anhydride (Figure 3.3) is a commonly used raw material with diacid functionality in unsaturated polyester resins. Those used in conservation are mostly styrenated polyester resins: the resin component and the curing resin give off styrene vapour.

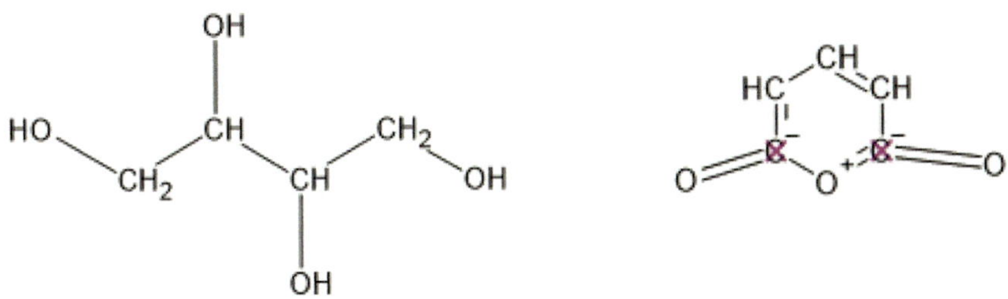

Erythritol
(a)

Maleic anhydride
(b)

Figure 3.3 Structural formulae for (a) erythritol and (b) maleic anhydride which can be reacted to form polyester resin.

Like epoxy resins, polyester resins have good cohesive strength and can therefore be used for casting missing components as well as joining fragments. They have been used to repair breaks in building stone and sculpture, to adhere metal dowels in place, and as clear casting resins for metallographic and paint samples. Just as for epoxy resins, their use as an adhesive is usually irreversible in practice, and if used in unsuitable situations their breaking strength might exceed that of a degraded object. Many examples are water-white, and not at all viscous when mixed which means that bubbles from mixing can all escape, which leads to the casting being of good quality and appearance. Some polyester resins can be much more readily ground and polished than epoxy resins.

Polyester resins act as allergic sensitisers for some individuals, and this may be more of an issue than for epoxy resins. Their tendency to release styrene as vapour is also a health and safety concern.

B3 Cyanoacrylates (Superglue®)

Cyanoacrylate adhesives were developed later than epoxy and polyester resins and more for the domestic user or the industrial user. They react with moisture and therefore set quickly in normal indoor environmental conditions: there is always a microscopic amount of water on the surfaces to be joined, and this promotes a rapid reaction to form a very strong adhesive bond with weak cohesive strength. Reversibility is a serious concern, and their requirement for clean, well-fitting surfaces poses definite limitations on use. One suggested application is for tacking glass fragments quickly into position by the minimum number of spots of cyanoacrylate glue, to support the joins while epoxy resin for example develops its maximum cohesive strength. Once the fragments are accurately in position, epoxy resin adhesive can be drawn into the break by capillary action to become the main adhesive joining the fragments together.

Cyanoacrylate adhesives are based on alkyl-2-cyano acrylate monomers. One of these, with a methyl group, is very similar to methyl methacrylate, which polymerises to form PMMA. The only difference is that the cyanoacrylate has a cyano (–C≡N) side-group instead of the methyl linked to the carbon backbone.

The adhesive is usually supplied in a container as a monomer combined with an acid that prevents the polymerisation reaction. If the acid is neutralised, then the monomer polymerises and becomes solid. The hydroxyl (–OH) groups found on most surfaces in the presence of water will neutralise the acid. On applying the monomer to a surface, the acid is removed and curing starts. If there are too few hydroxyl groups present a solid joint will not be formed. This will occur, for instance, if the surfaces are acidic, as happens with wood since this releases acetic acid, or if the atmosphere is too dry. The smaller the side-group, the stronger and stiffer the joint, and these products have high breaking strength. The larger the side-group, the slower is the curing time. Since the methyl monomer cures very rapidly it may be preferable to use the butyl monomer if the assembled parts have to be aligned or manipulated after contact. Although a strong bond is formed in a few seconds, complete curing takes a few hours, and during this time the joint can be removed using acetone. Their ageing characteristics have not been investigated for conservation use, so they should be used with caution, if at all.

C Melt-freeze joining

Although the vast majority of adhesives used in conservation fall into the two categories already discussed, there are times where neither solvent nor reaction adhesives are suitable. In those cases better results may be achieved by **melt-freeze adhesives**.

A melt-freeze adhesive is a material that is melted to allow a joint to be made whilst the adhesive is liquid; on cooling the adhesive re-solidifies. By definition, therefore, it has to be a thermoplastic. These materials provide the most rapid way of making a completed joint because they do not involve either evaporation of solvent or a chemical curing reaction.

Naturally occurring organic materials have been used as melt-freeze adhesives in conservation work for many centuries; examples are rosin (colophony), shellac, and beeswax. In ancient times sealing wax was made from beeswax, and in the last two centuries or so it has been made from a mixture of shellac and rosin mixed with turpentine or oil, and pigment. The turpentine is added as a plasticiser to increase the toughness; with age the turpentine diffuses out and evaporates and the wax becomes brittle. Synthetic waxes produced as by-products of the distillation of coal or refining of crude oil are also used; examples are paraffin wax (a straight-chain hydrocarbon derived from petrochemicals) and microcrystalline wax (a branched hydrocarbon).

Other advantages of the materials cited are that they melt at relatively low temperatures – and are therefore not likely to induce thermal damage in most objects – and they are relatively stable chemically. However, because the polymer molecules have few if any reactive (polar) side-groups, their adhesive and cohesive strength is relatively weak. In addition, when molten they have a high viscosity and so they are difficult to spread. They would be inherently unsuitable for an object that would encounter an uncontrolled and hot climate, since beeswax and some paraffin waxes have a melting point around 60°C.

Improved wax formulations exist commercially, mainly because of the use of waxed paper for wrapping bread and of waxed cartons for beverages. These developments provide a good example of how knowledge of the internal structure of a material can suggest alterations which will improve its properties. Paraffin wax and, as its name implies, microcrystalline wax are crystalline thermoplastics consisting of long hydrocarbon molecules. With no polar side-groups, intermolecular bonding is very weak and so the solid waxes are very soft. One way of improving the strength without affecting adversely other properties would be to add polythene, which itself is very similar to paraffin wax but with extremely long molecules. Unfortunately, it is not easily soluble in the waxes because of its crystallinity. However, ethylene-vinyl acetate co-polymers (known as EVA) *are* soluble. The EVAs are random co-polymers and essentially amorphous. As a result they are more compatible with paraffin and other waxes. A bonus obtained from such compositions is that the polar C=O side-groups contributed by vinyl acetate increase the adhesive bonding potential.

Another reason for describing how these materials work is that somewhat similar adhesives were developed for use in conservation (particularly of paintings and textiles) in the 1970s–1980s, specifically the BEVA® 371 adhesive based on EVA, mentioned earlier. The harmful solvents such as toluene used to thin them for use and the need to use applied heat from a spatula or hot-air blower, both during application and to reverse such repairs some decades later, should be done with appropriate PPE. Such materials are actually a variant of melt-freeze adhesives and are called **heat-seal adhesives**. The basic idea behind this type of heat-seal adhesive is to use a long-chain polymer (an ethylene-vinyl acetate co-polymer) for structural strength together with short-chain polymers (usually hydrocarbon resins) with a melting point in the range 80–100°C. On heating, the resin melts to form a free-flowing (low viscosity) liquid in which the co-polymer dissolves. It is used in this condition and on cooling reverts to a solid mass.

Solvent bonding is a related concept, used to join PMMA and other plastics, especially in an industrial context for the manufacture of display cases or transparent covers with no visible joins. Here, localised heating and dissolution of the pieces to be joined by the solvent allow them to be clamped together until cool, in effect creating the same material properties in the body of the PMMA and the joint itself.

C1 Soldering and brazing

Not all melt-freeze adhesives need be polymers. For a great many years melt-freezing has been used to join pieces of metal by the techniques known as *soldering*, *brazing*, and *welding*. In these instances the adhesive is a metal alloy which, in principle, behaves in a similar fashion to organic melt-freeze adhesives. Soldering, brazing, and welding are widespread metal-working techniques, and the science of joining by these means has significant implications for metals conservators. Soldering and brazing are similar in principle, so these will be considered before welding, which has some considerable differences.

Both of these processes are similar to the melt-freeze waxes that have just been discussed, in that pieces of a (metal) object are joined by an adhesive that has a lower melting point than the pieces to be joined. The main difference between soldering and brazing lies in the composition, and particularly the consequent melting points, of the adhesives that are used. Soldering generally refers to joining at lower temperatures than brazing. In both cases, the adhesive melts at a considerably lower temperature than the metal of the object.

In **soldering**, the adhesive (the solder), in the form of a wire, is melted (using a soldering iron) and then allowed to spread over the surfaces to be joined. The liquid solder interacts with the surfaces of the metal pieces to be joined and forms a very thin adhesive layer of *alloy*. The solder itself is a metal alloy, and bonding occurs as a result of the intermingling of atoms that occurs across the interface between the solder and the metal object. Random atomic movement is occurring rapidly in the hot liquid solder, but there is virtually no movement in the cold solid or interchange of atoms between the two materials. As thermal energy is transferred from the molten solder to the solid metal, the metal object warms up and so movement of atoms between it and the solder becomes easier. When the source of heat is removed, the solder cools and solidifies. Some of the atoms in the solder will have moved (diffused) into the metal object, and *vice versa*. Since the atoms in the solder and in the metal object are held together by primary metallic bonds, there is a continuity of primary bonding across the interface. Thus, in soldered joints, the major means of adhesion is primary chemical bonding. Although atoms from the solder only penetrate the object for a seemingly short distance (typically a quarter of a millimetre), a soldered joint is stronger than those obtained with solvent-adhesives and about as strong as those made with reaction adhesives. Furthermore, solder is a **ductile** alloy; that is, it can be plastically deformed. Consequently soldered joints are *tough* rather than brittle (see Chapter 4), which means that they are usually more robust and able to stand more severe impacts than joints made with organic adhesives. Soldered joints can be separated by heating, but the fragments will have been structurally altered by the process.

The wire for soldering is made from an alloy of (usually two) metals that has a suitable melting point. The most familiar solders are made from alloys of tin, and they melt at quite low temperatures, typically in the range 180–200°C. Historically the alloy would have included lead, but not today. They are used on low melting metals such as pewter or where higher temperatures would damage the object (if lead solder is already present on silver, high temperatures will cause total disintegration) or decoration (gilding, enamel, and so on).

Brazing is a similar process to soldering, indeed the two terms describe gradations of heat rather than different processes. A convention that could be used is that soldering is performed at temperatures below 300°C, brazing at temperatures above 450°C. Traditionally, brazing alloys are commonly split into two classes, high-temperature brazing and low-temperature brazing. Low-temperature brazing is sometimes called silver soldering or hard soldering. The brazing alloys are based on the silver–copper system with various other additions and have melting points in the range 600–900°C. They are commonly used for articles made of silver. High-temperature brazing is a particularly good joining method for brasses (alloys of copper and zinc) and bronzes (alloys of copper and tin) and can also be used for cast iron and wrought iron. The brazing alloys used are based on the copper–zinc system with a melting range of about 850–900°C.

In both soldering and brazing, the joining alloy can be melted at a point on the joint and is drawn into the joint by capillary action. In the light of what has been said in Chapter 1, you will appreciate that for capillary action to work effectively, an adhesive must wet the surfaces of the object to be joined. Since both the adhesive (solder or brazing alloy) and the surfaces to be joined are metallic, it may be supposed that there would be no problem in the molten alloy wetting the object. This is not the case however because, for most metals, the surfaces to be joined will not be clean on a microscopic scale: they are usually tarnished in some way. In general, because they are not particularly stable chemically, metals react with chemicals in the atmosphere. The predominant reaction is with atmospheric oxygen, which produces a layer of oxide on the surface of the metal. When heat is applied the rate of this chemical reaction increases, and the resultant surface film stops the molten adhesive wetting the metal and therefore causes a weak joint to be formed; the molten adhesive does not penetrate the joint, and the surface film inhibits the formation of an adhesive bond. For this reason, the oxides and other contaminants have to be removed, and this is done by using a special substance, called a **flux**, in conjunction with the adhesive. A flux is a material which, when molten, wets the tarnished metal and reacts chemically with the oxide or other contaminant to lay bare the clean underlying metal surface. The flux and the products of its reactions do not remain in the joint once the molten adhesive is introduced because they are much less dense and therefore float on the molten adhesive. Clearly, these features, plus the need for the flux to be molten at the soldering or brazing temperature, put stringent requirements on the choice of material for a flux.

Brazed joints are several times stronger than soldered joints and, when used on silver or copper alloys, of a strength comparable with that of the original metal. Structural changes in the two pieces once joined would be more pronounced – and separation of a brazed joint would be challenging in any case.

The relative strengths of different metal joining techniques can dictate which is/has been used in a particular case in conservation. For example, the handles of pewter and silver jugs and cups tend to come away from the main body of the object after many years of constant use. These may be re-attached by an organic adhesive, but the adhesive bond achieved may not be sufficient for the object to be lifted repeatedly by the handles alone, and the joint will probably fail. However, a lead solder joint on a pewter object, or a silver solder (brazed) joint on a silver object, should be sufficiently strong so that the handles will not come away from the object when supporting the whole of its weight. For strong joints in objects made of iron alloys or higher melting-point metals, an even higher temperature process, welding, would have to be used to facilitate handling as experienced in the service life of the object.

C2 Welding

There are two types of **welding** process; those that involve a combination of *pressure* and usually heat, but not sufficient to cause melting; and *fusion* welding which relies on partially melting the pieces to be joined, usually in conjunction with a filler *metal.* Welding is not exclusive to metals. For instance, many plastics can be joined by pressing them together when warm; the seams in polythene bags are formed in this way. Broken candles or wax seals can be repaired by melting the wax in the vicinity of the crack with a heated spatula. In both of these examples, adhesion is achieved by diffusion across the joint interface, and usually the original interface is completely destroyed in the process.

An important example of welding under pressure and heat is the process known as **hammer-forging** practised by blacksmiths. Here, two pieces of metal are welded by heating them and hammering them together while they are still hot. Some metals will weld even without heat being applied. Two clean flat pieces of lead, gold, or silver can be joined by simply hammering them cold. Again, the original interface disappears.

In **fusion welding**, the joint is heated intensely and locally, usually using a gas torch or an electric arc, and the join is formed by melting of the two surfaces, together with the 'filler' metal (if one is being used) to fill the gap between the pieces. It is similar to soldering and brazing in that the adhesive bond is due to the primary metallic bonding produced between the pieces as a result of inter-diffusion. The original interface disappears. Because of the very high local temperature involved, up to 3000°C, fusion welding is much more difficult to control than soldering and brazing, and it is a specialist skill. The major problem is rapid oxidation of the parent metal. This and other technical demands of the process have led to the development of a variety of techniques, but it is not appropriate to discuss them here. Welded joints are very strong, and when well made are as strong as the parent metal. However, welding raises issues if it is used in conservation, rather than the fabrication of the original industrial object or steel sculpture. The most important one is that the melting of an alloy can lead to compositional changes; an important example of this is the **dezincification** of brass in which zinc is lost from the alloy because it is a relatively volatile element.

Perhaps the most serious problem arises from the **heat-affected zone** that is produced in an object that has been welded. This is the volume of material around the site of a weld in which the microstructure of the material is irreversibly changed. Typically, it can extend up to one centimetre from the region which was melted. It is not appropriate in this text to discuss in any detail the nature of the changes that can occur in the heat-affected zone, but you should know why such a zone is important: cracks would develop preferentially along its perimeter, and microstructures such as these can be analysed to determine the source of the material and the various mechanical and heat treatments it has undergone. In other words, by studying its microstructure, you can deduce the technology used to produce an object made of metal. In the heat-affected zone of a weld this information is destroyed.

Before leaving the subject of soldering, brazing, and welding, some other effects of heat on materials should be mentioned. These can be significant in conservation work and can arise when using melt-freeze adhesives.

C3 Melt-freeze adhesives and the effects of heat

The application of heat, no matter how small the temperature rise it produces, can cause problems. Even the heat from an organic wax that melts at about 60–80°C can lead to some objects losing moisture or to internal stresses that could be deleterious to fragile objects. Of course, at the higher temperature involved in soldering and brazing, and especially welding, the internal stresses are more severe. Fortunately, many of the metals used in the objects that a conservator would meet are usually sufficiently ductile to withstand these stresses. However, fusion welding can embrittle an object through the absorption of chemicals from the environment.

The internal stresses set up by the rapid heating and cooling of a metal object that has a coating (such as enamel or a corrosion crust) can lead to damage. For a given rise in temperature, the thermal expansion of metals is considerably greater than that of other inorganic compounds such as oxides, carbonates, and so on, which constitute typical coatings. Also, such compounds are usually brittle. Thus when a metal with a coating is heated, the metal expands more than the coating which means that the metal exerts a pull on it. A tensile stress is set up in the coating, which may break and spall off in flakes. For this reason it is inadvisable to use soldering, brazing, or welding on metal objects that are coated in this way. Apart from this problem the increase in temperature may produce colour changes in a patina, for instance in those found on old bronze.

4 The strength and stiffness of materials

From Chapter 1, it is clear that the strength of a joint depends very much on the bonding at the interfaces between the adhesive and the broken surfaces of the object. This depends on the magnitude of the secondary forces between the adhesive and the object and on the potential number of bonding points. But there can be complications: if the coating of adhesive is too thick, the strength of the joint depends on the strength of the solid adhesive itself rather than on the interfaces. The coating has to be extremely thin, just a few molecular layers, before the bulk strength of the adhesive can be ignored. From a practical point of view, the strength of the joint will be enhanced by a very close fit of the broken pieces. This enables the chemical bonding and mechanical keying to be at a maximum and reduces the effect of the bulk strength of the adhesive to a minimum.

Another important characteristic of a joint which is related to strength but, as you will see, utterly different from it, is its stiffness – or conversely, its floppiness. A joint between pieces of metals, glass, or pottery, for instance, needs to be stiff and ideally as stiff as the object itself; joints in paper, leather, or textiles, on the other hand, should be as flexible as the object's materials. There are of course cases where a flexible joint is an advantage with a stiff material, for example, adhesives which have flexibility could allow repaired window glass in buildings to flex in strong winds.

The words 'strength' and 'stiffness', together with other words we associate with the response of something when it is pushed or pulled, or knocked or put under pressure, such as 'force', 'stress', 'strain', 'brittle', and 'tough', are part of everyday vocabulary, where they don't have a universally agreed meaning, and may be used metaphorically and not literally. However, scientific models that explain how materials respond when subjected to loads require strict definition of each of these words, and in order to understand the behaviour of an adhesive in these terms, you need to be clear about their precise meanings. This is the subject of the next three sections.

A Stress and strength

Any push or pull exerted on an object is an example of a *force*. Engines create forces that drive things, you create a force on a glass when you lift it to drink or on

DOI: 10.4324/9781003389217-4

a suitcase when you roll it over a rough street, but less so over the smooth flooring of a station, and so on. A pulling force that separates two components is called a **tensile force** and a force that pushes them together is a **compressive force**. The *weight* of an object also produces a compressive force on the surface upon which it rests, and stone or brick in a wall exert a compressive force on the blocks lower down in the structure.

If you simply measured the strength of a joint as the force needed to break it (the **breaking load**), the result would only tell us something about that particular joint. Because the dimensions of the joint are important, the result would not give any information on, for instance, the strength of the bond between adhesive and object, or the strength of the cohesive bonds within the adhesive itself or within the material being joined. To get this information the effect of size needs to be eliminated from the measurement of the breaking load. Picture two rods of the same material where the cross-sectional area of one is double that of the other. If the two rods are pulled in tension, the thicker rod will support double the load that the thinner rod can take before it breaks. This was recognised a long time ago but it was not appreciated until very much later that if you divide the load needed to break a rod by its cross-sectional area, you get the same answer for rods of the same material no matter how thick they are. By doing this calculation, the effect of size is eliminated. The load on a piece of material divided by its cross-sectional area is called the **stress** in the material. When an object, such as a rod made of a particular material, breaks, the stress acting in it is the same irrespective of its cross-sectional area, and it is this stress that is called the **strength** of the material. Now look at this idea more formally.

Let's return to Figure 1.3, which showed three types of joint. Figure 4.1 illustrates the ways different forces act on each joint. Compressive forces were used when the joint was being supported or clamped until the adhesive set. Tensile forces show how the joint would be separated for repositioning during its adhesion, and also what would be done to reverse the joint at a later date. The purple arrows show the twisting, or application of a **shear force**, that the joint might encounter during handling and object movement. They have been scaled to show what one can readily feel about the relative importance of the threats posed by tensile and shear forces applied to each type of joint, whether accidentally in moving the object or deliberately when trying to reverse the joint. It is simplest to suppose that the tensile force is spread evenly over the whole of these joints. The stress is then calculated simply by dividing the force acting on the material by its cross-sectional area:

stress = force/area.

The same situation would occur for any and every cross-section. Notice that the stress acts in opposing directions because in order to put a load on the material it has to be pulled at each end. The stress created by a tensile force is called a tensile stress and that created by a compressive force, a compressive stress. Since stress acts at all points on a cross-sectional area of an object, it is basically a measure of the force by which the atoms or molecules in the material are being pulled apart or

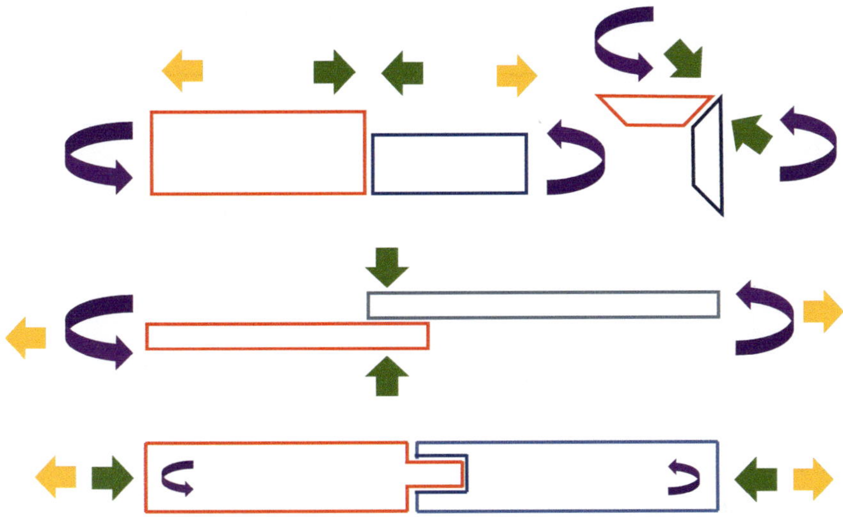

Figure 4.1 Schematic of forces acting on joints: compressive forces in green, tensile forces in yellow, and shear forces in purple, roughly scaled. Top left: simple butt joint, affected by shear. Top right: mitre corner (of a frame), also affected by shear. Middle: a simple lap joint, affected by shear. Bottom: a simple form of mortise (component on the left) and tenon joint with every cut forming a right angle, less affected by shear.

pushed together. Consequently when the tensile or compressive force on a material reaches the stress at which some of its internal bonds are broken, the material begins to break. The maximum tensile stress a material can support before breaking is called the **ultimate tensile strength (UTS)** of the material; you will frequently see it referred to simply as **tensile strength** or as its **breaking strength**, a term used in Chapter 3 to describe the same idea. Similarly, the maximum stress a piece of material can sustain in compression without breaking is called the **ultimate compressive strength**. (As you will see in Section 4C, these definitions need to be modified slightly for materials that flow under load.)

These concepts are described and illustrated very clearly, with conservation examples, in Wei (2021: 8–34), along with some simple calculations to show how practically useful such engineering concepts can be when selecting one adhesive over another.

Measurements of tensile strength provide a useful way of comparing different materials. Such measurements are made in specially designed tensile testing machines, using standardised sizes and shapes of test pieces, also discussed and illustrated in Wei (2021). Table 4.1 shows the approximate tensile strength of a range of materials. (The symbol MNm^{-2} in Table 4.1 stands for mega-newtons per square metre which is an internationally recognised unit for stress,[1] also known as mega-pascals.) Conservation research employing tensile testing often uses modified sample shapes and protocols, because conservation materials and degraded

Table 4.1 The approximate UTS of a range of materials. Some values taken from Wei (2021).

Material	*Relative strength*	*Approximate UTS (MNm^{-2})*
timber (across grain)	extremely weak	5
house brick		5
concrete without rebars		5
sedimentary stones like sandstone		5–30
tanned leather	weak	40
window glass		30–50
epoxy resin		50
polyester resin		50
phenol formaldehyde		50
porcelain		70
timber (along grain)	fairly strong	50–120
copper		140
cast iron		70–140
wrought iron		100–300
some polymers and plastic, when new		30–200
some glass and ceramics		50–500
human hair	strong	200
silk fibre		350
mild steel		400
bronzes and brasses		100–600
nylon thread	extremely strong	1000
high-tensile steel wire		2000

materials in objects have low UTS compared to the usual test subjects, so the methods have to be modified, including the way the samples are gripped in the machine, to give any distinction among their UTS values. A single measurement of UTS is not very useful: it is necessary to measure a number of examples and then work with average values and statistics, so clearly surrogate materials and mock-ups have to be used.

There is an enormous variation in the tensile strengths of different materials. This variation reflects the types of bonding and internal structural arrangements that occur in materials. For instance, note the huge difference in tensile strength for timber across and along the grain. Timber is a fibrous material and has a grain, as any carpenter or mediaeval maker of panels would know, because long molecules of cellulose are packed very closely together in bundles; the bundles resemble drinking straws packed in a box (Figure 4.2(a)). There is far less contact, and therefore less bonding, between molecules in neighbouring bundles than there is between molecules within a bundle. (This is also true of the molecules in drinking straws). So it is not surprising that timber is much stronger when pulled along the grain.

There is a very important caution about the data in Table 4.1. The data are for typical samples of a given material that have been prepared under controlled conditions and of standardised shape and size. Other samples tested in other conditions could give values somewhat different from those quoted, and this is particularly

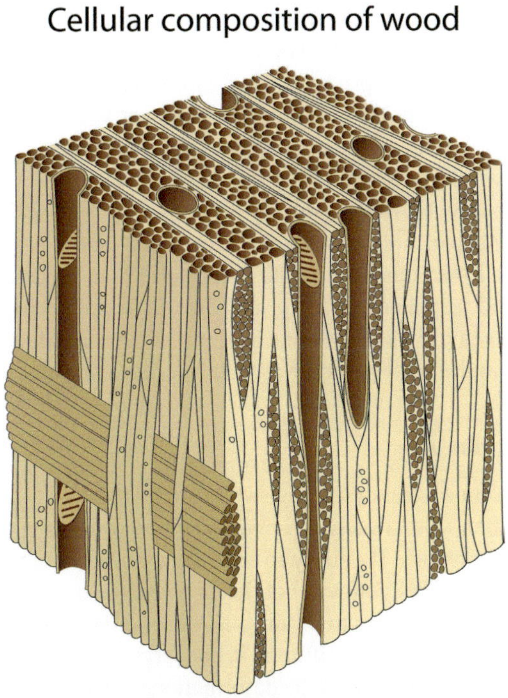

(a)

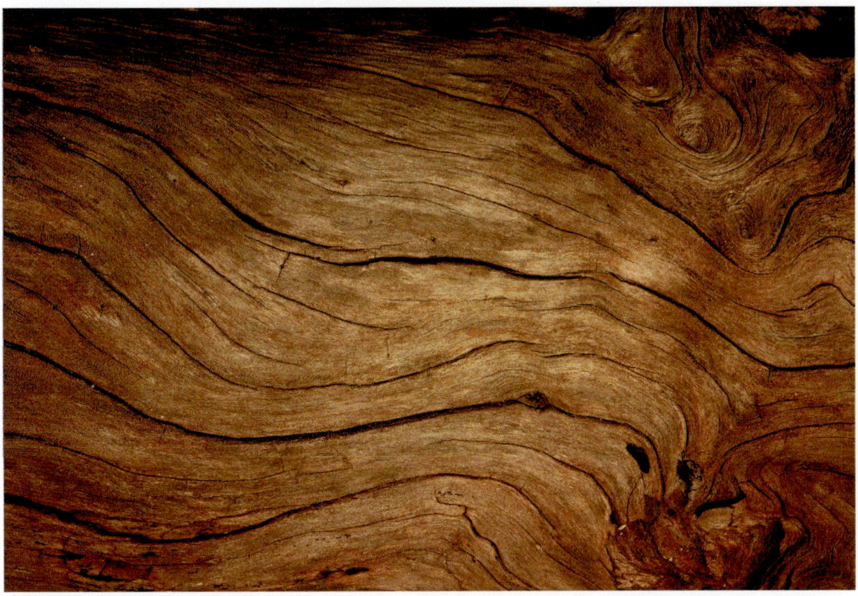

(b)

Figure 4.2 (a) The cellular structure of wood. (b) Cracks travelling along the grain.

Images: Shutterstock.

likely with the objects you handle. Most of the properties of a material, especially tensile strength, are not immutable. The internal constitution of a material depends on its history – how it was produced and subsequently treated. A sample with cracks (Figure 4.2(b)) or holes in it will have a lower tensile strength than would be expected for a perfect sample made from the same material (see Section 4D), and similarly so would those that have corroded, deteriorated over time, or been degraded in some way. However, bearing these limitations in mind, such data can be useful. For instance, notice that the 'strong' adhesives epoxy and polyester resin have a much lower tensile strength when solid than metals, but that they are stronger than wood tested across the grain. This raises the possibility that a joint made between pieces of wood with one of these adhesives could break within the wood rather than in the joint (which was stated earlier without the explanation).

B Strain and stiffness

All solids change their shape (deform) when subjected to a weight or any other kind of force. When you stretch a rubber band or squeeze a pencil eraser, the change of shape is obvious. If you try to squeeze a piece of glass, pottery, metal, or wood between your fingers you will not detect a change of shape because a much greater force is needed in order to produce an observable change. These materials appear to have a greater intrinsic stiffness. When large objects made from these stiff materials are subject to large forces the deformations *can be* observable; trees sway in the wind, aeroplane wings deflect in travel, and the tops of tall skyscrapers move a metre and more in a strong wind.

It is clear that some objects are stiffer than others, that is, they are more resistant than others to changing their shape under an applied force. A branch is stiffer than a twig; they have the same intrinsic **stiffness** (they are made of the same material) but differ in dimensions. You have to get rid of the effects of geometry to look at the material alone in order to study these phenomena in a scientific way. To understand why some materials are intrinsically stiffer than others you need to consider the bonds that hold atoms and molecules together in a solid. Envisage that the millions of bonds between atoms or molecules in a material behave like minute springs. When a material is stretched or compressed each of these bonds is stretched or compressed too, and the overall change of shape of the material is the result of the many very small changes at a microscopic level.

The intrinsic stiffness of a material is a measure of how much the atoms or molecules in a material resist being pulled apart or pushed together by an applied force; the stiffer the inter-atomic 'springs', the stiffer is the material. To obtain a more precise definition of the stiffness of a material, a way of measuring the amount of deformation (change of shape) produced is needed that is independent of the size of an object. The answer lies in the concept of **strain**.

Demonstration

Find a reasonably 'stretchy' elastic band and put marks on it 2 and 4 cm apart. Then stretch the rubber band alongside a steel measuring tape. You will find that the 4 cm length has stretched to twice as long as the 2 cm length. The rubber band has been stretched uniformly along its length. You can check this for yourself by putting lots of marks close together on the strip and observing what happens when the band is stretched. If the change in length is divided by the original length the degree of stretch is found to be the same for both the 2 cm length and the 4 cm length. It is called the strain. Expressing it as a formula gives:

stress = change in length/original length.

From this expression you can see that for a certain stress, the greater the amount of strain produced (the greater the percentage deformation) the less stiff is the material.

Strain is sometimes expressed as a percentage; for instance, an extension of 1 mm in an original length of 100 mm may be expressed as either 0.01 or 1%. In the prior example, the strain in the rubber band is 1/2 or 50%. When a pencil eraser is squeezed or trees and skyscrapers sway in a strong wind, the strain produced in each case is reversible under normal conditions. On letting go, the eraser returns to its original shape, and the tree and skyscraper return to their upright position when the wind stops blowing. This reversible deformation is called **elastic deformation**, as opposed to the permanent, irreversible **(plastic) deformation** that can be produced in some materials.

The stiffness of the material is the resistance it offers to being deformed elastically, and its stiffness can be measured as the stress needed to produce a specified amount of elastic strain. Notice that it is being defined in terms of stress and strain; that is, the size of the sample has been taken into account, and so the result is a direct measure of an intrinsic property of the material concerned.

Correctly, the stiffness of a material is called the **elastic modulus** (*modulus* is the Latin for 'a measure', and so it takes the correct Latin plural, namely *moduli*). For a given material there are several different elastic moduli, depending how evenly and how fast the load is applied. So far we have only looked at tensile and compressive loads; for these cases the elastic modulus is called **Young's modulus**. Table 4.2 shows approximate values for Young's modulus for a number of materials.

Table 4.2 Ranges of flexibility and Young's modulus for a selection of materials.

Material	*Relative flexibility*	*Young's modulus (MNm^{-2})*
natural rubber	very flexible	7
polythene	fairly flexible	200
wood (across grain)		500
epoxy resin	rigid	4500
phenol formaldehyde		7000
wood (along grain)	very rigid	10,000–15,000
glass		70,000
stone		10,000–80,000
bronzes and brasses		70,000
iron and steel		200,000
diamond	extremely rigid	1,200,000

The units for measuring Young's modulus are the same as those for stress, MNm^{-2}. This is because strain is a number (and therefore it has no units), so dividing stress by strain alters the value but not the units. Looking at the values from rubber to diamond, it is not surprising that you cannot squeeze a piece of pottery or glass between your fingers as you can a pencil eraser. Notice also that epoxy resin and phenol formaldehyde have a much lower Young's modulus than metals, glass, stone, and wood along the grain; so, they would make glued joints with these materials that are more flexible than the materials they are joining.

It is important to remember that just as different metals have different values of Young's modulus, so will different formulations of epoxy resin. It is possible for you to adjust the physical properties of a resin, as will be seen later.

In many instances the forces which act on a joint will not be those of compression or tension. The example of *appliqué* art, where wood, metal, or glass, say, are stuck to a vertical surface, would be a case where different forces operate. This different system of loading has already been met in Chapter 1C when you parted two wetted glass slides by sliding one over the other. In this case you would be applying a shear force too. Tensile forces acting on an object have a common line of action, and so have compressive forces.

One reason why shear forces are important is that in many materials the force needed to make atoms or molecules slide over one another is less than that needed to pull them apart. This provides the possibility of irreversible **plastic deformation** in which a material undergoes a permanent change of shape. It is a characteristic of most metals and of thermoplastics and glasses above their glass transition temperature. This phenomenon is looked at in more detail in the next section.

In practice, many objects are subjected to forces which are neither purely tensile nor purely compressive, but combinations of the two. Bending, the most important example of this, arises when a tensile or compressive load is applied to an object at a point a distance away from the points at which the object is supported. A floor joist spans the walls in a house and when you stand on it, it bends, and in an old building with wooden joists one might even be aware of this; when you lift a saucepan by the handle, the weight of the contents is acting at a distance from your hand, so the handle bends but not perceptibly, and so on. You could determine the type of deformation produced in bending, using a piece of modelling clay or similar (Plasticine® or PlayDoh®) or a pencil eraser.

C Plastic deformation

Plastic deformation is permanent irreversible deformation. It is what happens when you squeeze a piece of modelling clay or warm beeswax; they do not return to their original shape when you release the force on them.

Plastic deformation occurs in addition to elastic deformation; all solids deform elastically to some extent, but some deform plastically as well. You can demonstrate this for yourself with articles such as a polythene bag. If you release it after pulling it slightly, it will return to its original shape; this is elastic deformation. Increase the pull and you will find that beyond a certain point you start to deform the article permanently; this is plastic deformation. When you do this, notice that when you are pulling hard enough to produce plastic deformation and then let it go, the bag or grip relaxes a little; part of the deformation is still elastic (reversible). The ability of a material to be deformed plastically is very important. It is one of the reasons why metals are so useful; for example, car bodies can be pressed to shape rather than being machined from a solid block of steel, a plumber can bend a copper pipe but could not do the same thing with a ceramic pipe, and so on.

More generally, a material that can undergo plastic deformation is usually more difficult to break: drop a glass mug and it will break, drop a metal mug and the worst thing that can happen is that it is dented. Glass is said to be **brittle** and a metal **tough**. In the next section we explore the meaning of each of these terms.

D Brittleness and toughness

If all the pieces of a broken pottery plate or glass window are recovered, they will often fit back together to form the original shape, although occasionally a vessel may not fit back together perfectly due to internal stresses from its manufacture or firing. The pot is then said to be 'sprung'. This is a common characteristic of **brittle** materials; they show virtually no plastic (permanent) deformation before breaking. Stone, ceramics, and cast iron are other examples of brittle materials, as are thermosetting adhesives, coatings, and consolidants. They have a useful advantage in conservation work, especially if they are also weak, because they are more readily removed (mechanically) from an object.

Brittle materials are particularly susceptible to fracture under tensile stresses, but are generally stronger in compression. This can be useful. If you want to break a sheet of glass in a rectangular shape the technique is to scratch a line on the surface and then bend the glass 'against' the line. It should break neatly along the scratch. When a sheet of glass is bent like this, its top surface is stretched, in tension, and the bottom surface is shortened, in compression. The technique does not work if the scratch is on the compression surface.

The scratch in the surface is the key to understanding what happens. The tensile stress across the top surface of the sheet is the same at all points except near the scratch, which is the cause of a much higher local stress at its tip. Such high local stresses are called **stress concentrations** and they occur at notches, scratches, cracks, and holes in materials that are otherwise subject to uniform stress. (Sharp bends or corners in an object can also be sources of stress concentration.) The sharper the defect, the greater is the stress concentration. Even if a material looks smooth, there will be minute surface defects that can act as sources of stress concentration, such as those on the surface of the fragment of broken glass shown in Figure 1.1. The tensile stresses acting in a material are, in effect, magnified around the source of a stress concentration.

As Figure 4.3(a) illustrates, a tensile stress tends to open up a crack and cause it to grow further into the material. A compressive stress acts to close the crack and inhibits its growth effectively, as shown in see Figure 4.3(b). Plastic deformation

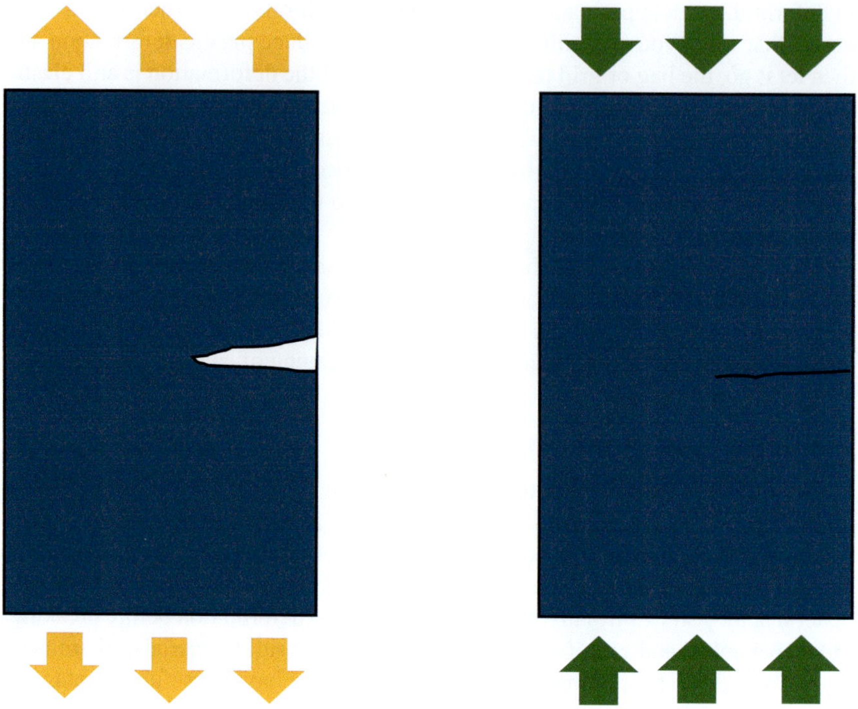

Figure 4.3 Schematic of a block of glass (for example) with a crack in it. (a) Uniform tensile stress makes the crack grow. (b) Uniform compressive stress closes up the crack, and is less damaging both immediately and in the longer term.

makes a material tough because it can occur at the tip of a crack or tear and has a blunting effect, thereby reducing the stress concentration at the tip.

The ability of materials such as metals and thermoplastics above their T_g (see Chapter 3) to deform plastically makes them **tough**, which means that they are relatively difficult to break. A polythene bag may look flimsy but you would find it difficult to break one (or tear one, which seems a better word here).

Some materials which do not exhibit plastic deformation can be tough because they possess another intrinsic crack-stopping mechanism. Wood and leather are examples. If you break a piece of wood you can fit the broken bits back together to form the original shape. They are brittle (that is, they do not exhibit plastic deformation), yet they are difficult to break. This apparent anomaly lies in the internal structure of these and similar materials. Figure 4.2(a) shows the fibrous cell structure of wood, which on a larger scale you see as the grain of the wood. The structural component of the cells – cellulose molecules – are aligned along the length of the fibres. Cracks that attempt to grow across the grain in wood are simple diverted along the grain. Figure 4.2(b) illustrates this for a piece of wood loaded along the grain. Using this explanation, how would you expect wood to behave when loaded

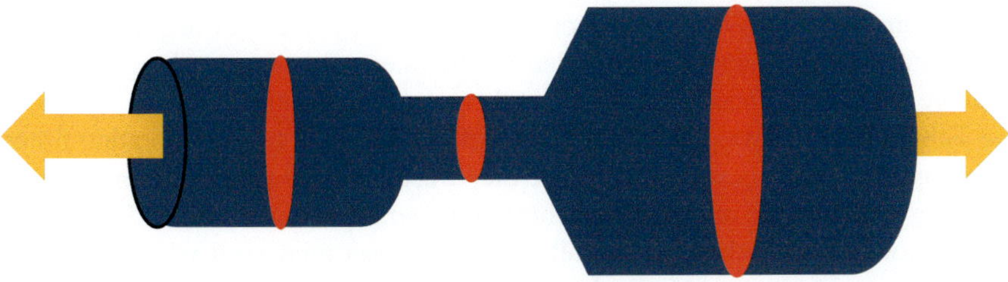

Figure 4.4 Schematic of an object of uniform composition, subject to tensile stress. The red cross-sections represent points A (on the left), B, and C (on the right).

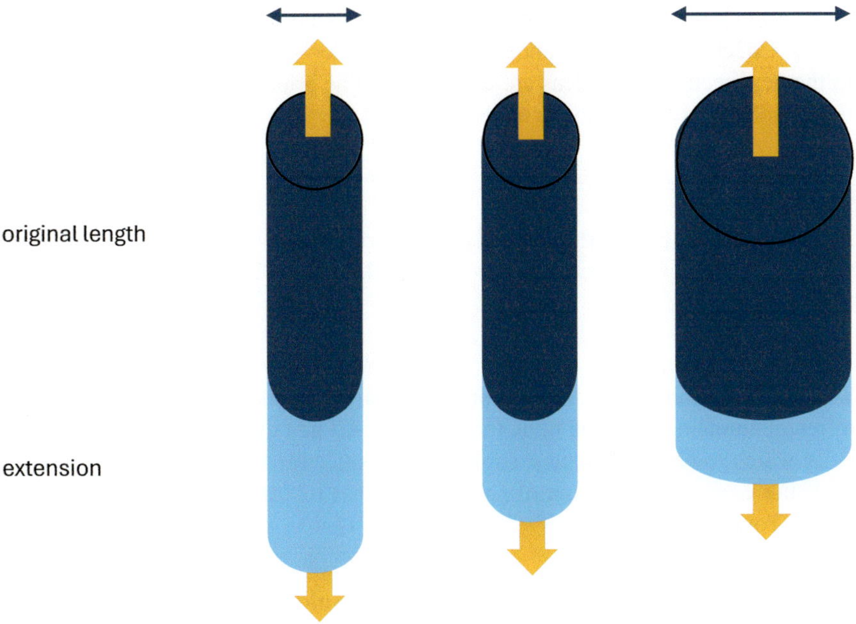

Figure 4.5 Three rods made from three different materials (shown in dark blue), each one of a uniform composition, under tensile stress that has caused each to stretch to a different degree (the light blue portion). (a) and (b) are 1 unit in diameter and (c) is 2 units in diameter.

across the grain? It should be weak and break easily, and, in fact the tensile strength of wood is about 20 times greater *along* the grain than across the grain. Most plants and other natural materials, including bone, horn and ivory, are fibrous and therefore tough. Of course nature ensures that the fibres are aligned in the likely direction of the largest stresses they will encounter in life, and it pays when using these materials (or imitating them) to copy nature. Craftspeople would always have done this.

The material commonly known as glass reinforced polyester resin is toughened by crack-stopping mechanisms similar in principle to those in wood: the glass fibres are there to carry out this function. It has found a wide variety of structural uses, from the bodies of cars to replicas of museum objects. In this material the crack-stopping mechanism is provided by the interfaces between the glass fibres that are used to reinforce the (polyester) thermosetting plastic.

This idea of making a material more difficult to break by incorporating fibres was appreciated many centuries before the effect could be accounted for. The ancient Egyptians made mud bricks incorporating straw, and in medieval Europe the plaster for surfacing the indoor walls of houses was reinforced with horse hair. In the nineteenth century, perfectly serviceable and cheap chairs were made from *papier mâché* (pulped paper mixed with a starch-based binder); the cellulose fibres in the paper were the strengthening and toughening agents.

In the last three sections a number of concepts have been introduced and terms used in the description and explanation of how materials respond to mechanical forces. Check that you have understood them by tackling the following exercises. In the next section these ideas are used to examine briefly the general characteristics of adhesives and the joints made with them.

Exercise 4.1

Assume that the (peculiarly) shaped rod in Figure 4.4 is being pulled under a certain tensile force. On which cross-section, *A*, *B*, or *C*, will the stress be greatest?

Exercise 4.2

The rods in Figure 4.5(a–c) are of different materials, and have been stretched by different amounts under the same tensile force. All had the same original length.

In which rod has the greatest strain been produced?

If the cross-sectional area of (c) is twice that of (a), which material has the higher Young's modulus?

E Adhesives, adhesive joints, and their mechanical behaviour

In earlier sections the way that an adhesive joint responds to a load has simply been referred to as the 'strength' of a joint. You can now appreciate that a number of different material characteristics play a role in determining how well a joint responds to loads. You have been introduced to the formal language needed to consider these characteristics, in particular to definitions of force, stress, strain, ultimate tensile strength, Young's modulus, and terms like tough, brittle, elastic, and plastic. This section looks at adhesives and joints using what you have learnt about these terms.

First, there is a general point that applies to the use of adhesives in most conservation work. The requirement is for an adhesive that will produce a joint that is strong enough to withstand the stresses to which the object is normally exposed, which include supporting its own weight. In the event of a failure, it is a better option by far for the joint to break either within the adhesive itself (which is preferable because less damaging) or at an interface between the adhesive and the adherend. In other words, the joint should not be so strong that the break occurs within the body of the object. In addition, there is conflict between the need for the adhesive bonding to be strong enough to do its job and to last, yet weak enough to be broken down safely should there be a need to take the joint apart. Consider some of the factors that influence the effectiveness of an adhesive joint – after all, whatever the required strength of the joint, along with its other characteristics, the ultimate aim is for its properties to be predictable just after it has set, and into the future also.

The loads to which an object is subjected: those generated when it is handled or used, or accidentally knocked, or the weight of the parts, elicit a variety of responses in a joint through the stresses they generate. It is the stress generated in a material (stress = force/area) that produces fracture. So, in general, for a given load on a joint, the lower the stress, the less likely the joint is to cause problems. This argument leads to two significant conclusions. Firstly, the greater the area of contact between the pieces to be joined, the lower is the stress generated in the joint by a given load; the load is spread over a greater area. This reinforces the point made in Chapter 2, Section B, that the adhesive should cover the whole of the available break surface. It also emphasises that a joint with a large cross-sectional area will be stronger than one with a smaller area. This applies if all other things are equal, which brings us to the second conclusion. For a given load, the stress is magnified in the vicinity of any source of stress concentration. Stress concentrations occur in any region where there is a change of cross-section or a flaw of some kind. Consider some examples. Fortunately, many of the joints needed in the repair of broken objects simply involve butt jointing two pieces of the same cross-sectional area. This is essentially what happens when you stick together the pieces of a pot or plate, pieces of flat glass, or regularly shaped pieces of metal. These are all examples of butt jointing, which is illustrated in Figure 4.6(a). Problems with stress concentrations arise with joints of the type shown in Figures 4.6(b) and 4.6(c).

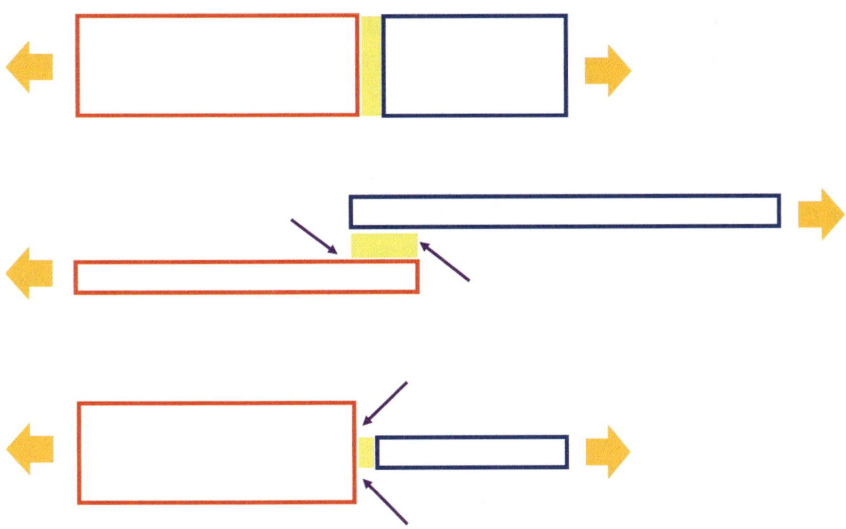

Figure 4.6 Schematic of (a) butt join of two thick pieces of material; (b) lap join; (c) butt join of a small to a larger piece. All with adhesive (the pale yellow block) joining them. The purple arrows show stress concentrations.

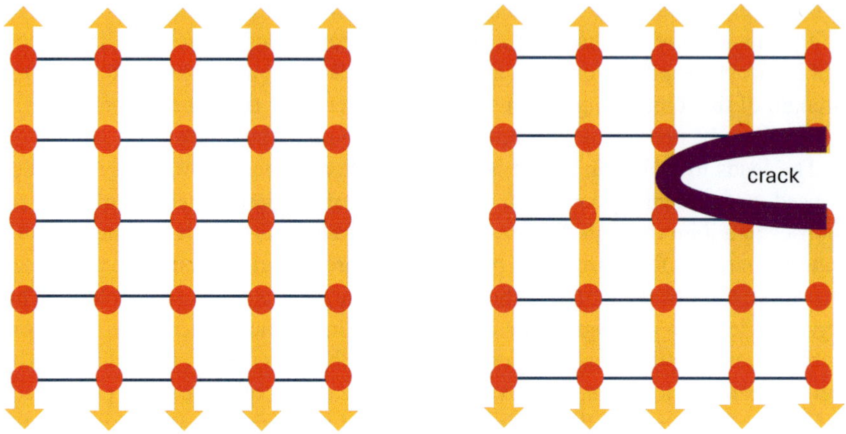

Figure 4.7 (a) Schematic of a block of material at the atomic scale. (b) The crack from the side, as in Figure 4.2(a), visualised at the atomic scale. Purple indicates the stress concentration.

Examples of the type of joint shown in Figure 4.6(b) may not 'feel' common, although a badly repaired object might look like this before the old repair is taken apart, but those of the type shown in Figure 4.6(c) are common. The joining of handles to pots and cups is one example.

Some idea of the origin of a stress concentration can be gained by considering a simple model of what happens on an atomic scale at the tip of a crack. The grid of atoms depicted in red in Figure 4.7(a) can be visualised as a simple chain of atoms under tension due to an externally applied load. In this idealised crystal each chain is uniformly stressed. Now, what happens if there is a crack in the structure? In effect, the crack in the position shown in Figure 4.7(b) is created by breaking some neighbouring bonds. This means that some of the chains of atoms are broken and, if the crystal is to remain whole, the load in these chains has to be taken by those that go round the end of the crack. The load in the broken chains is then concentrated on the one, or the few, atomic bonds at the tip of the crack, shown in purple. In fact, the stress concentration is greatest just here, at the tip of a scratch, crack, or notch. This is why a paper or tinfoil will tear easily once a small slit is torn/cut in the edge and why glass breaks along the line previously scratched on its surface. It is also the reason why shrinking of adhesive is undesirable. As the glue line recedes from the surface, a depression, and thus a stress concentration, is formed (Figure 4.8). Cracks will appear here first, and will propagate in the areas of highest stress.

The example of handles is a good illustration of how, in the past, potters appreciated the problem and tried to overcome it, without a physical explanation of the forces involved. Look at the handles on cups, vases, or jugs, and you will see that the join between the handle and the body of the object is smoothly curved, and that the handle itself often also broadens out where it meets the cup or vessel: the greater the area of the section the lower is the stress, and the larger the radius of

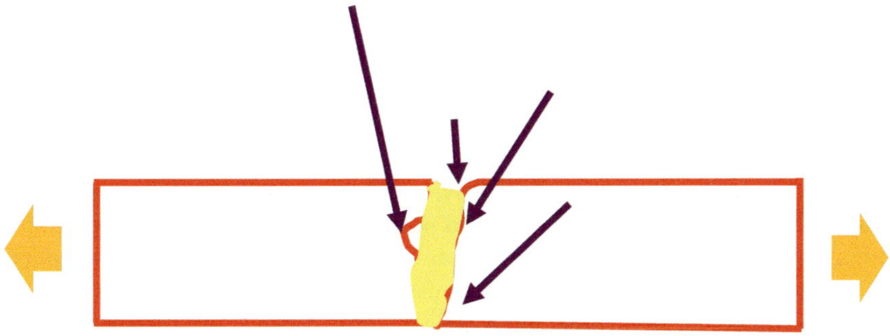

Figure 4.8 Schematic of a rough break, not adequately filled with adhesive, shown in pale yellow. Here, the purple arrows are roughly scaled to show areas of stress concentration when the repair is pulled apart, i.e. experiences a tensile force.

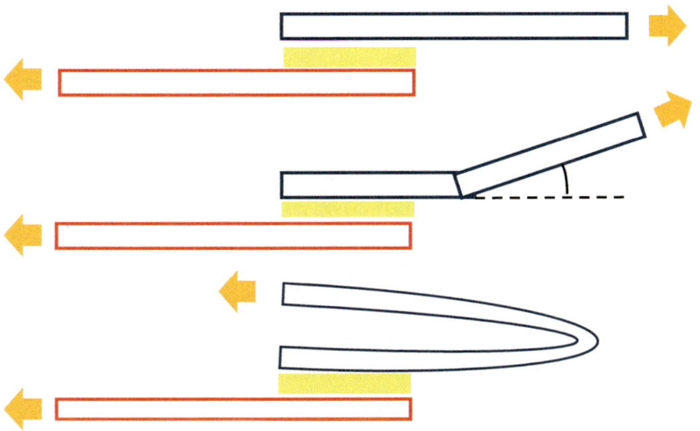

Figure 4.9 (a) Schematic of tensile force applied as the joint with adhesive (shown in pale yellow) is pulled apart. (b) Shearing as well as tensile force is one form of peel testing. (c) The most extreme set-up for peel testing.

curvature, the lower is the stress concentration. If the change of section at the join is abrupt, then the object is inherently fragile.

The important point here is that it is primarily the stress concentrations that determine how a joint behaves under load, and the larger the radius of curvature the smaller is the effect of the stress concentration. In a lap joint shown in Figure 4.6(b), the stress concentration at the arrowed points dominate how the joint responds to a load. The distance over which the pieces overlap is largely irrelevant compared with the width of the joint (its depth perpendicular to the page).

Incidentally, the lap joint configuration illustrated in Figure 4.9(c) is the basis of the **peel or stripping strength test** that is often used to measure the effectiveness of a flexible coating on a rigid surface, for use as paint films for instance.

This kind of test has also been used for assessing lining adhesives for paintings or non-stitched supports for fragile textiles.

In Figure 4.8, the block represents the rough or weathered break to be joined, while the adhesive shown in yellow is not coating the break surface very well (perhaps it was too viscous to flow over it and wet it). If the joint is being pulled apart it is under tensile stress, and worse, if it is being pulled apart at an angle it is also subject to shear. The non-adhered portions experience different degrees of stress concentration (easier to imagine than to sketch!). The more shear is applied, the more this resembles a **peel test**. The result of such a test is a value of the **peel strength** (measured as a force, not a stress) and it depends on, for instance, the elastic moduli of the rigid surface and the flexible material and the peel angle. Interpretation of this measure is complex, and since peel strength is measured as a force (not stress), it can only be useful as a comparative test, that is, in measuring the behaviour of different adhesives using test pieces of exactly the same size and shape and in exactly the same geometrical configuration. Obviously these tests have to be standardised or different people's results could not be compared (Figure 4.9, Kollaard and Wei 2023).

On a smaller scale, flaws in a joint, such as bubbles in the solid adhesive or cracks in the adhesive or at the interfaces of the joint, act as points of major stress concentrations; any source of a crack is a source of trouble – see Figure 4.8, where these points of weakness could occur from inadequate preparation of the joint, particularly if the surfaces of the object are not covered completely with adhesive and/or not pressed together firmly, or from a flake of detached object only loosely attached to the break surface. Under the tensile load shown in Figure 4.8 both flaws would tend to grow as cracks because the force acts to open them up.

Having dealt with some of the factors that affect the strength of a joint, what about its stiffness requirements? Ideally, the elastic modulus (or simply elasticity) of the solid adhesive should be the same as that of the object material. In some cases, this is an obvious requirement – for instance, joining pieces of a stiff object with a flexible glue might produce undesirable distortion of the object. There is another, less obvious, reason. If part of an object has a different stiffness (elastic modulus) from other parts, any stresses that arise are not evenly distributed through the object; the stress may be concentrated in particular regions. To illustrate this point, consider a veneered piece of wood. Suppose that it is subjected to a tensile load that stretches it by a certain amount (Figure 4.10). Young's modulus = stress/strain, and since the

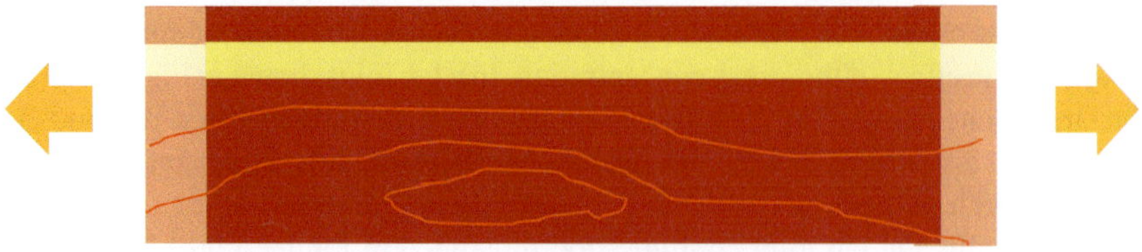

Figure 4.10 Schematic of wood veneer attached with adhesive (shown in pale yellow) to a block of wood, lengthened under tensile stress.

strain produced in the adhesive and the veneer is the same, if the Young's modulus of the veneer is the greater, there will be the stress produced in it. Correspondingly, if the adhesive has the greater Young's modulus, a greater stress will be generated within it. In either case the shear stress created is acting to break the adhesive bonds at the interfaces between the adhesive and the bonded surfaces. In the conservation of such a structure, a stiffer adhesive is preferred because it usually means that, if failure occurs, the adhesive breaks rather than the object.

Before concluding this section on the mechanical behaviour of adhesives and adhesive joints, one further point about stress and strength should be introduced. This is that stresses can be generated within the body of an object in the absence of an externally applied load. These **internal stresses** as they are called can exceed the strength of the material and therefore lead to fracture. Any process that leads to part of an object undergoing elastic deformation relative to those parts to which it is attached will set up internal stresses. An important example of such a process in adhesive joints is **thermal expansion**. Most materials expand when they are heated, and some more than others. Thermal expansion is similar to stiffness in that its value depends on the magnitude and character of the bonding forces between the atoms or molecules in a material. The stronger the bonding, the smaller is the thermal expansion – the smaller the distance the atoms and molecules move apart. For example, polymers bonded by secondary forces expand more than materials, such as metals and ceramics, that are held together by primary bonds. Thus, for a given rise in temperature, many adhesives will expand more than materials they are joining. When this happens it is equivalent to a greater strain occurring in the adhesive than in the adherend, with the result that internal stresses are created. This effect can be of practical significance in a number of ways. Imagine coating a thin strip of material with molten beeswax, or some other melt-freeze adhesive. When the adhesive cools, solidifies, and continues to cool down, it contracts progressively. (In other words it gets shorter.) Provided that the beeswax remains stuck to the thin strip, the strip is forced to bend to the shape of the coating; the beeswax is under compression and the strip under tension. Of course, the bending occurs in this example because the strip is unconstrained. If the strip were a bulky object, or, say, held at its ends, the internal stresses would still be there. Similar effects are produced by solvent adhesives. In these cases the contraction is not due to cooling but to the shrinkage associated with the evaporation of the solvent. Related effects arise if a solid adhesive expands by absorbing moisture. You can see that the behaviour of an adhesive is complex, and that the problems of, say, solvent evaporation or water absorption can affect the stability of an adhesive joint. This is the subject of the next chapter.

Note

1 The Système Internationale (SI) unit of force is the newton (N); 1 N is about the weight of an apple, 100 grams. Stress (force/area) is measured in newtons per square metre (N/m^2 or Nm^{-2}) or pascals (Pa). Pressure is measured in the same units. Since one newton is such a small force and the square metre is a large area, stress is often given in mega (million) newtons per square metre (MNm^{-2}).

5 How long will it last?

How long an adhesive joint lasts obviously depends on what degradation processes are taking place and how fast. Physical or chemical changes occur in the adhesive or adherend that make the joint incapable of coping with the mechanical loads acting on it or with the internal stresses that are often generated by those changes. Paradoxically, perhaps, many of the processes that lead to degradation may be harnessed by you when you wish to remove an adhesive joint from an object. Consequently, exploring these processes, in this chapter, has a double purpose.

Clearly, except for circumstances in which an adhesive joint is subjected to an abnormally high load (when the object is dropped, for instance), it should be permanent, provided that it is chemically stable and in a controlled harmless environment. However, neither of these conditions is guaranteed in practice, over the long term.

The elements that make up a metal or alloy are not usually a stable state compared with their natural state as minerals (gold is the exception); and they react with chemical compounds in the atmosphere, such as oxygen, sulphur dioxide, hydrogen sulphide, carbon dioxide, and water. This is why, for instance, rust forms on many iron alloys and silver becomes tarnished. Similarly, both synthetic organic polymers and materials made from natural organic polymers (such as paper and leather) oxidise in air, leading to discoloration and embrittlement. The presence of water can have quite different effects on polymeric adhesives; it can be absorbed and produce swelling, and hence internal stresses (in, for instance, adhesives based on nylon), and it can react chemically with the polymer to produce both structural weakening and undesirable products (such as acetic acid from cellulose acetate and polyvinyl acetate). Finally, embrittlement can be produced in an adhesive that loses a component by slow evaporation. This applies particularly to solvent adhesives which retain some of the solvent and to adhesives that have constituents added, such as plasticisers.

In this chapter the main physical changes that can lead to degradation are explored. Chemical changes are dealt with next and, finally, the degradation of joints in practice. The distinction between chemical and physical changes was made in *An Introduction to Materials and Chemistry*. Chemical changes are processes in which the primary bonds between atoms are re-organised to produce new chemical compounds.

DOI: 10.4324/9781003389217-5

A Physical changes

Physical changes usually involve a change of volume due to the gain or loss of material, or the rearrangement of the existing atoms or molecules. The extent and rate of volume change depend primarily on two phenomena: diffusion and solubility.

A1 Diffusion

You know from earlier books in this series that the atoms and molecules in a solid are never entirely static. Even at very low temperatures atoms vibrate and the bonds in molecules stretch and contract. In fact, atoms and molecules are sometimes able to move through the bulk of a solid; small molecules are more mobile than large ones as they can squeeze through the gaps between other molecules. This movement is called **diffusion** and it plays an important, often crucial, role in many processes. For instance, it determines the rate at which a solvent is removed from a joint when a solvent-based adhesive is used. The small solvent molecules have to diffuse through the tangle of much larger and less mobile polymer molecules. As a reaction-type adhesive sets, the monomers have to pass through the rapidly forming polymer network to reach the appropriate reaction site and so the rate of diffusion may determine the rate of setting. One of the main reasons for applying a coating to an object is to protect it by providing a barrier between the object and harmful chemicals. Clearly, in such cases, the most effective coatings are those which minimise diffusion through them of the harmful species; more will be said about this in Chapter 7.

This section firstly deals with how diffusion occurs in a crystalline solid composed of atoms bound together by primary bonds, such as a metal or a ceramic, and then moves on to discuss diffusion in polymers, which has some rather different characteristics but which can be understood in terms of the scientific models introduced here and earlier.

Diffusion in inorganic crystals

The atoms in a crystal form a regular arrangement as a result of a balance in the chemical bonds acting between them. Also, atoms are in a constant state of motion which determines the kinetic energy they possess. The higher the temperature, the greater their kinetic energy, and it is this kinetic energy that drives the process of diffusion. However, it does not drive atoms in any particular direction: diffusion is, in fact, the result of the haphazard wanderings of many individual atoms. Figure 5.1 shows (in 2D) three stages in the diffusion that occurs when two different metals are placed in intimate contact. This is illustrating what we saw happening in the soldering process in Chapter 3. In this very simplified example just a very small part of a plane that cuts through the pair of metals will be considered. If the metals are elements of 'blue' type and 'red' type, then at first all the atoms at one side of the interface will be blue and all those on the other side will be red atoms. We must assume that a blue atom will be as stable when completely surrounded by red atoms as it is surrounded by blue atoms, and that a red atom is equally stable with either as its nearest neighbours. That is, we assume that the blue and red metals are completely soluble in one another, as, for instance, copper and nickel are. Now consider a blue atom in Figure 5.1(a) at the interface between the two metals. It can move in any one of six directions – up, down, to the left, to the right, into the page, and out from the page. (In this model a move refers to an exchange with a neighbouring atom.) The only one of these possibilities that makes a difference to the overall distribution of atoms is a move to the right, and so of all the moves made by all of the atoms, only a fraction lead to an obvious compositional change. In Figure 5.1(b) a few blue and red atoms have diffused across the interface, and the distribution is being smoothed out. As diffusion continues, the distribution of blue atoms will be progressively evened out until the distribution represented by Figure 5.1(c) is reached. Here the interface has actually disappeared – the solid on one side is indistinguishable from that on the other. To leftwards and rightwards but not shown, diffusion may not yet have occurred. Although diffusion will still continue, it will make no difference to the overall distribution of the atoms. Statistically, for every blue atom moving to the right, say, there is a red one moving to the left.

The grossly simplified model of diffusion described here has the sole purpose of showing how the random migration of atoms produces a change in composition. It ignores a number of factors that are important in real materials, and, in particular, it assumes that diffusion of blue atoms into the 'red' metal is as easy, or as difficult, as that of red atoms in the blue metal. Since the magnitude of the chemical bonding between blue atoms, between red atoms, and between a red and a blue atom will all be different, the relative ease of diffusion of the two species will be different. A different scenario would be a weak solution of a solvent-based adhesive, which would therefore have a lot of solvent in, applied to a plastic. The solvent, with a lower molecular weight than the plasticiser in the plastic, would diffuse in and cause swelling because the plasticiser would diffuse out more slowly. (This is one reason why applying a solvent-based adhesive to any plastic might well represent a bad treatment decision.)

It has been found by experiment that the rate of diffusion is proportional to the **composition gradient**. Composition gradient describes the way composition changes with distance. In Figure 5.1(a) the gradient is very great across the interface as there is a total change in composition from all blue to all red in a very small distance. The gradient is beginning to level out in Figure 5.1(b) as the distance between mostly blue and mostly red increases. The composition gradient in

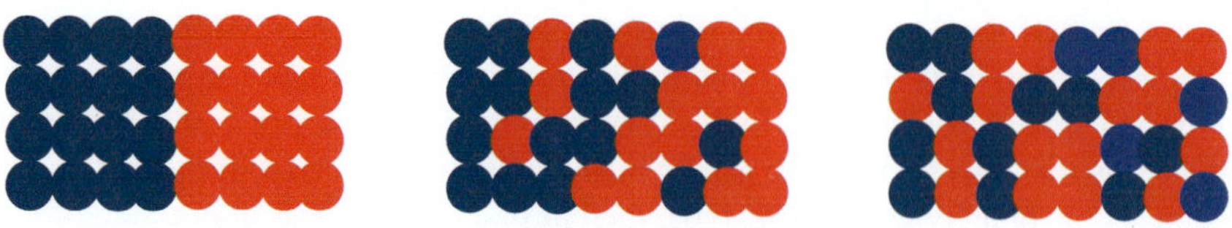

Figure 5.1 Schematic of diffusion between two different metals, whose atoms are shown in blue and red. (a) Metals in contact. (b) Diffusion is occurring across the interface. (c) The whole depicted zone is a mixture of the two metals.

Figure 5.1(c) is zero as there is by now uniform composition throughout the solid. The **rate of diffusion** can be thought of in terms of the net number of atoms that migrate across a given plane within the solid (such as the interface in Figure 5.1) in a given time. This number can be represented by the letter J.

The proportional relationship between the rate of diffusion J and composition gradient, C, can be expressed as a very simple equation:

$$J = D\,C$$

This means that the rate J is equal to some number, D, multiplied by C, the composition gradient. D is called the **diffusion coefficient** (or **diffusivity**). It is constant for one material diffusing through another material at a particular temperature. The units often used are m^2/sec or m^2/hour.

The relationship described by this equation is called **Fick's Law** (although it would be expressed more conventionally with C described mathematically as a rate that could then slow down over time), but it is not the name or the symbols that are important, it is the simple relationship between rate and concentration (composition is the concentration of one substance, A, in another, B). Values for the diffusion coefficient can be determined experimentally, for instance, by surrounding a material with a radioactive isotope of the required chemical element and monitoring the progress of the isotope through the material. Table 5.1 gives you an idea of the sort of values obtained for different materials under different conditions; they are very approximate, but give an idea of the huge ranges found in practice.

You will notice that there is an incredibly large difference between the values. Unfortunately the diffusion coefficient does not give a direct measure of the distance over which diffusion occurs. However, since diffusion is the overall result of the random movement of individual atoms, the random movement can be analysed to give the average behaviour of an atom. In a certain time, call it t, an average atom migrates a distance which can be calculated as

$$\sqrt{(Dt)}$$

from its starting point, the square root of D multiplied by t. So, the distance A is that over which the composition can change substantially in a time, t, due to diffusion.

Table 5.1 Rough values for the diffusion coefficient and the time for diffusion to occur over 1 mm, for different types of materials.

Material and diffusion species	*D*	*Time for average atom/molecule to diffuse 1mm*
single atoms in metal at 25°C	10^{-25}	10^{23} seconds (will never happen)
single atoms in metal (near its melting point)	10^{-8}	10^{6} seconds = about 12 days
single atoms in a liquid metal	10^{-4}	10^{2} seconds = 100 seconds
small molecules such as oxygen, in a polymer	10^{9}	10^{-22} seconds (almost instant)

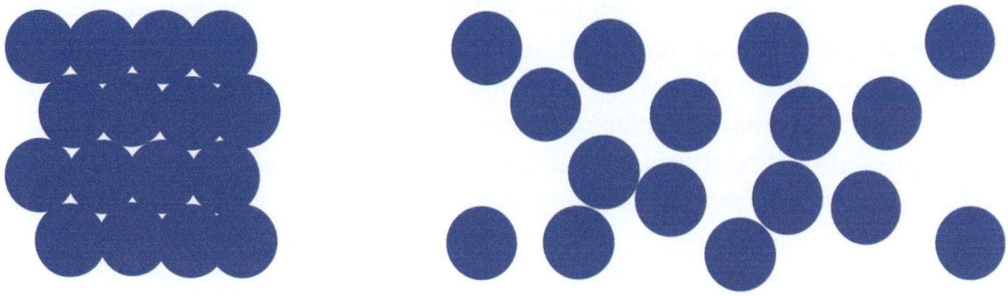

(a) (b)

Figure 5.2 (a) Schematic of a solid, crystalline metal. (b) Liquid metal.

To give you an idea of what this means, Table 5.1 includes the time it takes, on average, for each of the diffusing species to travel 1 mm. The range is enormous. A detailed explanation of why the difference between the diffusion of atoms in metals at ordinary temperatures and of small molecules in polymers is so great is complex and beyond the scope of this book. However, qualitatively, the reasons are not difficult to see.

Consider first the difference between a solid and a liquid metal. In a solid metal the atoms are strongly bound together by primary bonds and packed tightly together in a crystalline array; the arrangement is illustrated schematically in Figure 5.2(a) – which is in fact more realistic packing than was shown in Figure 5.1, because the arrangement in Figure 5.2(b) is a lower energy configuration. In the liquid shown in Figure 5.2(b), the atoms are more loosely packed: the liquid takes up more volume than the solid did.

More energy is required by an atom in order to move into a neighbouring site in a crystal than in the liquid, mainly because it has to disturb its neighbouring atoms much more. The (relatively) very fast rates of diffusion that can occur in metal crystals near their melting point simply illustrates the dramatic effect of the greater thermal energy available at higher temperatures in aiding a diffusing atom to overcome its bonding forces.

Diffusion in polymeric materials

Diffusion of the molecules in a thermoplastic material differs from that of atoms in a crystal in two important ways. Firstly, the polymer molecules are very large and have a complex structure, and so very little movement past other molecules occurs. Secondly, since the use of such materials is usually limited to temperatures below about 100°C, the available thermal energy is much less than that for metals at higher temperatures. Consequently, diffusion of the polymeric molecules in a thermoplastic below its glass transition temperature (see Chapter 2, Section D2) can be assumed to be negligible. Of course, in thermosetting plastics the rigid network polymer molecule will not undergo diffusion either. However, smaller molecules can diffuse within and through a polymeric material and often very readily. This is because of the relative openness of the molecular structure – particularly in amorphous polymers – and because, in thermoplastics, the molecules are only held together by weak secondary forces.

In a thermoset the polymeric network is relatively open and diffusion will be easy, depending on the size of the diffusing molecule and the size of the spaces in the network. Clearly, the more frequently cross-links occur along the polymer molecule, the smaller are the gaps in the network, and consequently the smaller are the molecules that can readily diffuse through the structure. Similarly, amorphous thermoplastics permit more rapid diffusion than do crystalline polymers. Diffusion through an amorphous thermoplastic varies sharply, depending on whether the temperature is above or below the glass transition temperature, T_g. When the temperature is above T_g, the mobility of the polymer chains allows paths to open up for the diffusing molecules; at temperatures below T_g, the structure is rigid and consequently diffusion is more difficult – it requires the aid of more thermal energy, than at higher temperatures. This is shown in Table 5.2, where polymers in different states are shown to have very different rates of permeability to oxygen molecules. **Permeability** is a complex phenomenon involving diffusion and solubility factors. A small polar molecule will permeate a polymer with polar side-groups more readily than a non-polar molecule. Not only can the polar molecule pass through gaps in the network because of its small size, it can do so because it is compatible with (effectively soluble in) the polymer: the saying 'like dissolves like' can be extended to 'like permeates like'. This is demonstrated by the data in Table 5.3 which shows the relative permeability of several polymers to water vapour.

Poly(vinylidene) chloride is built up from $CH_2 = CCl_2$ monomer units and as a co-polymer with vinyl chloride is the basis of waterproof films such as Saran®. Polythene has no polar groups on its hydrocarbon chains, whereas polyesters have carbonyl groups at regular intervals in their structure. Cellulose acetate, formed by treating cellulose with acetic anhydride ($(CH_3CO)_2O$) has the polar acetate group and some remaining –OH groups which are compatible with the water molecules.

Table 5.2 Relative permeability to oxygen at 30°C for different polymers.

Polymer	*State*	*Relative permeability to oxygen at 30°C*
Poly(ethylene terephthalate) (PET)	crystalline	1
Polystyrene (PS)	amorphous	50
Natural rubber	rubbery	1000

Table 5.3 Relative permeability to water at 90% RH and 25°C.

Polymer	*Relative permeability to water at 90% RH and 25°C*
Poly(vinylidene chloride) (PVDC)	1
Polythene (PE)	10
Polyester film e.g. Melinex®/Mylar®	100
Cellulose acetate (CA)	5000

An obvious consequence of diffusion is the *swelling* that occurs when molecules migrate into and occupy the gaps in a polymeric structure, particularly in thermosets and amorphous thermoplastics. Before developing this point further, it will be useful to explore briefly polymer solvents and solubility.

A2 Polymer solvents and solubility

The **solubility of polymers** is of particular significance in conservation work. Firstly, many adhesives, coatings, and consolidants that have traditionally been used need to be in solution in order for them to be applied effectively. Secondly, dissolution or at least swelling using a solvent is an important means of removing old joints that were made with such adhesives in the past.

When a long-chain (thermoplastic) polymeric material dissolves in a solvent, the primary bonds holding the atoms together in each molecule remain intact; each polymer molecule retains its identity. This is demonstrated by the fact that the polymer molecules can be recovered from a solution by, say, evaporation – rather as salt can be recovered from brine. To form a polymer solution molecules of the solvent penetrate the polymeric structure and separate the polymer molecules from one another. This is depicted schematically in Figure 5.3. For a polymeric material to form such a solution, the secondary bonds between adjacent polymer molecules must be broken. This is favoured when the bonds between the solvent molecules and the polymer molecules are equal to, or stronger than, those between polymer molecules. One consequence of this is that solvent molecules with polar side-groups are likely to be able to penetrate and ultimately dissolve polymer molecules with polar side-groups. Thus polar liquids such as water, alcohols, and ketones have been widely used as solvents for adhesives; cellulose nitrate dissolves in ketones or esters, for instance, and poly(vinyl alcohol) (PVAl) is soluble in water.

Many thermoplastics have limited solubility in solvents in the sense that only small amounts of solvent molecules enter and permeate the polymer material. As discussed in Section D of Chapter 2 such solvents act as plasticisers; they increase

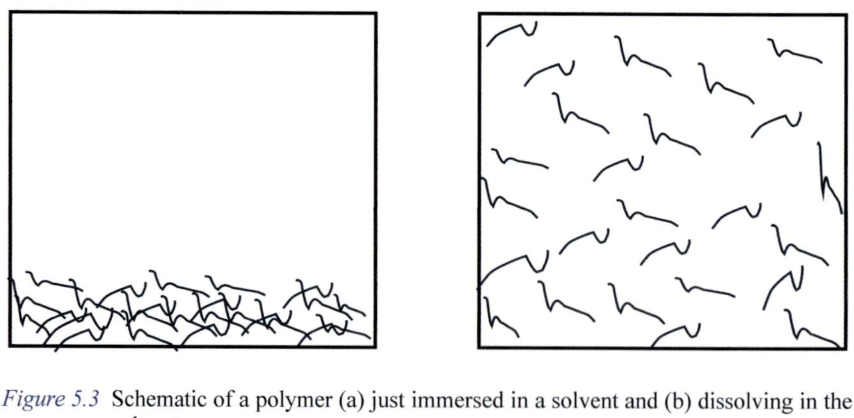

Figure 5.3 Schematic of a polymer (a) just immersed in a solvent and (b) dissolving in the solvent.

the flow (plastic deformation) capability of the material, reduce its stiffness, and in effect reduce its T_g. The absorption of such molecules into the gaps between the polymer molecules usually causes the material to swell. As you would expect, because they have a more open structure, amorphous long-chain polymers permit more rapid diffusion than do highly crystalline polymers. Since swelling requires diffusion of the relevant molecule, materials of high crystallinity have a more limited tendency to swell.

If an established adhesive joint, coating, or consolidant absorbs solvent molecules, and water is the obvious and most important example, the processes of plasticisation and **swelling** can lead to degradation. This is looked at in Section C. (Remember that the same processes can also be the means by which old conservation materials are removed from an object.)

Cross-linked (thermosetting) polymers cannot be dissolved in the sense described here. Since they consist of a network in which the atoms are held together by primary bonds, individual polymer chains cannot be separated by solvent molecules. However, chemically compatible liquids may penetrate them to some extent and produce swelling which may eventually soften or crack the material. Again, this can lead to degradation and also be a means of removing old materials from an object.

B Chemical change

A chemical change, as stated at the beginning of this chapter, involves a chemical reaction, that is a process in which the primary bonds between atoms are rearranged to produce new chemical compounds. Generally speaking, the parent molecules in the natural and synthetic polymeric materials used in conservation work are relatively stable chemically. However, there is one particularly important exception to this statement: their susceptibility to oxidation. It should not be surprising that most organic polymers suffer from oxidation; after all they are usually flammable (combustion being rapid oxidation). This is one reason why very few plastics used in an engineering or fabrication context are useful above about 100°C. However, for most applications at ordinary temperatures the rate of oxidation is tolerably slow or it can be controlled, but, unfortunately, in conservation work the required time scales are relatively long. In adhesives, coatings, and consolidants, oxidation can cause degradation and hence embrittlement or discoloration.

A useful illustration of the embrittlement produced by oxidation is provided by natural rubber and its synthetic counterparts (these are usually called **elastomers**). Oxygen atoms, like sulphur atoms, can form two covalent bonds and can therefore provide similar cross-links to those made by sulphur in the vulcanisation of rubber. This is what happens as an ageing process in the rubber used in, for example, vehicle tyres. Progressively, oxygen atoms form more cross-links, and, as you know from Section C in Chapter 2, the greater the number of cross-links the harder and more rigid is the material. In effect, the rubber becomes more like a thermosetting plastic and, in a tyre, the rubber is unable to undergo the large elastic deformation required every time the wheel revolves. As a result the rubber begins to crack at the surface; you can often see such cracks (known as 'perishing') on the walls of old tyres.

In a similar way, when other polymeric materials are embrittled by oxidation, it is usually due to the creation of cross-links (the other main reason for embrittlement is due to the loss of plasticiser or solvent by evaporation, see Chapter 5, Section C). It is becoming clear that oxidation does not normally occur by direct reaction between oxygen and a polymer chain. It is thought that under normal temperatures and conditions certain highly reactive impurities incorporated in the material during its synthesis or processing are normally involved. The end product is that a C–C or C–H bond is attacked. Details of the mechanisms are beyond the level of this text, but from this simple description it is clear that polymeric materials used in conservation should be as pure as possible. Another reason for this is that discoloration is often due to the oxidation of impurities, as in phenol formaldehyde for instance, which is always yellow or brown. In some polymeric material oxidation is accelerated by ultraviolet light. This is called **photo-oxidation**. Again the mechanisms are complex and diverse but it seems that the breaking of C–C or C–H bonds does not usually occur; the C=O bond appears to be much more susceptible, which means that polyesters, polyurethanes, and polyamides (such as nylon) are particularly prone to photo-oxidation. Note, however, acrylics are generally much more stable, so the whole picture is definitely not simple.

Oxidation can be inhibited, but not arrested completely, by incorporating **anti-oxidants** in the formulation of a plastic. These normally work by either reacting preferentially with oxygen that diffuses into the polymer or with the active impurities that may be present.

To summarise: in chemical terms the more stable polymeric materials are, the less they are likely to undergo chemical change. In other words, reactive molecular structures and components should be avoided. From this and earlier chapters in this text, you can start to recognise a number of points about the structure of polymers that could encourage chemical stability:

- Avoiding materials containing unsaturated polymer chains – those with a C=C bond such as rubbers and unsaturated polyesters.
- Avoiding materials containing polymers with highly reactive side-groups such as C–Cl.
- (Unless objects are shielded from ultraviolet light) avoiding the use of polymeric materials containing C=O bonds.
- Using materials that are as pure as possible.

These points are generally true but there are notable individual exceptions, such as some stable and very useful polyesters and acrylics.

C Further thoughts on the degradation of adhesive joints

Since the function of an adhesive joint is to hold pieces of an object together, that is, to resist loads attempting to pull the pieces apart, the joint will have failed when it can no longer withstand these loads. However, in conservation work the requirements are often more demanding. For instance, a joint may be judged unacceptable, although still structurally sound, because it has become distorted or discoloured. The main reasons for discoloration were discussed in the previous section, and so here only degradation processes that involve the mechanical behaviour of a joint, that is its responses to mechanical loads, will be considered.

Adhesives applied in liquid form end up as a solid in the completed joint – the reasons behind these requirements were examined in Chapter 1 Section C. In a melt-freeze adhesive the liquid cools and freezes to a solid of the same composition. Solution adhesives and reaction adhesives are not so straightforward. To produce a solid joint, the solvent in the former case, and the water, or other products, of the polymerisation process in the latter, have to be removed. The molecules involved are usually small and most of them diffuse to the surface and evaporate during the time the adhesive is hardening. During this stage the adhesive is in a paste-like condition with a fairly open structure, so diffusion is easy. However, some of the solvent or polymerisation product will remain in the solid joint and diffuse out over a long period.

The important point about these changes from liquid to solid and the loss of solvent from the solid is that they are always accompanied by a shrinkage of the adhesive. This **shrinkage** is important not only because it can lead to distortion of the object – particularly if the joint does not have an even thickness – but also because it produces internal stresses. The consequences of this are explored with an idealised adhesive joint illustrated in Figure 5.4; to demonstrate the effects, the thickness and shrinkage of the adhesive have been exaggerated. On shrinking, the molecules in the adhesive move closer together and therefore pull the pieces of the object being joined closer together. If the other ends of the pieces are fixed, as they would be in many cases by interlocking fragments, say replacing the handle on a pot, a piece in a broken plate and so on, contraction of the adhesive creates a pull (a compressive force) on the joint or the pieces being joined. If these internal stresses reach values that exceed the strength of the adhesive, or the bonding at the interfaces, or of the pieces of the object, fracture will occur. These stresses can be important, particularly for very fragile materials, such as aged textiles or paper, and in joints that have become weak as a result of chemical change. This is not the only effect. Shrinkage will occur in all directions within the adhesive, which means that internal *shear* stresses are also created (Figure 5.4). These stresses will act directly on the adhesive joint. The material on either side of the joint is deformed elastically in shear, which means that the original chemical bonds across the interface will be distorted.

This is an important effect because regions of elastic strain are more prone to attack by chemicals in the environment than those that are not strained. Such attack could progressively remove the bonds at the interface between the adhesive and the adherend with consequent weakening of the joint. Under these conditions the interface could be, for example, the preferred site for oxidation or a preferred route for diffusion of solvent molecules such as water. The latter process of degradation has been found to be particularly important for adhesives used under normal weathering conditions. Epoxy resins are an example.

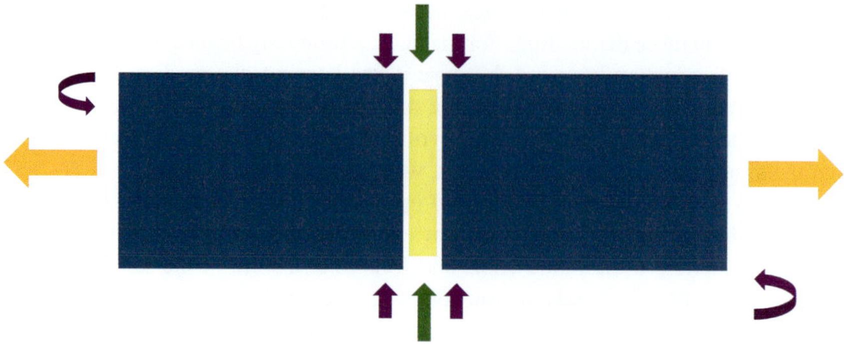

Figure 5.4 Schematic of shrinkage in a set joint made with a solvent-based adhesive, in effect imposing tensile stress (yellow arrow). The adhesive (pale yellow) experiences compression (green arrows). Applied shear (purple) by displacement or twisting acts on the points of stress concentration where the adhesive forms a step down from the fragments being joined.

The 'drying out' of an adhesive joint, with the consequent creation of internal stresses and an increased susceptibility to chemical attack, is one of the main factors that can ultimately lead to the complete breakdown of a joint. Clearly, the more an adhesive shrinks as it ages, the more likely it is to degrade and eventually break down. So, for instance, the amount of solvent used in a solvent adhesive should be kept to a minimum.

Another very important aspect of the loss of solvent during ageing is that in many adhesives (cellulose nitrate, poly(vinyl acetate), and some acrylics are examples) small amounts of residual solvent molecules act as a plasticiser. If the solvent-free adhesive is a weak brittle material, the progressive loss of solvent will lead to embrittlement. This is an inevitable process that will be made worse by chemical changes such as oxidation and **photo-degradation**.

So far only the creation of internal stresses as a result of shrinkage has been looked at. Stresses can also be created by the reverse process, that is swelling (expansion). If objects on which the conservation treatments have involved joining, coating, or consolidation are kept under mild environmental conditions, the problems associated with the absorption of liquids and consequent swelling are seldom significant. Although moisture from a normal atmosphere may diffuse into, say, a joint, it will be in insufficient amounts for the internal stresses produced by any swelling to be of consequence; as indicated earlier, the disruption of adhesive bonds by water molecules is likely to be much more important. Swelling, however, is a crucial problem in the restoration of waterlogged objects made from timber, paper, and similar materials. The internal stresses produced by swelling on the one hand, and by the contraction produced in drying timber and longer in waterlogged soil on the other, can readily lead to cracking, warping, and failure. These effects are produced, for instance, when the relative humidity fluctuates.

A further undesirable effect of the ageing of adhesives (as well as coatings and consolidants) is that some give off damaging chemicals. These chemicals may be a gas that is harmful to people and/or is damaging to other materials. For instance, degrading cellulose acetate and poly(vinyl acetate) can give off respectively acetic acid, which accelerates the corrosion of lead and other metals, and hydrochloric acid which could damage other materials in a composite object. Both of these acids would accelerate the degradation of any polymers that undergo hydrolysis: in the case of degrading cellulosed acetate, this positive feedback is very harmful if the object is enclosed with its self-generated acidic vapour. Many of the solvents long used in solvent adhesives are harmful, or even toxic in a few cases.

This discussion has stated fairly emphatically how difficult it is to choose a satisfactory adhesive. Selection is made more difficult by the fact that the technical requirements of commercially available adhesives are often less demanding than those of conservation, particularly with respect to discoloration. If you do consider commercial adhesives, try to use those that have been well-tried and characterised, and if experimenting with new ones it is best to use those that contain few impurities or additives, stable anti-oxidants, and a minimum of solvent or plasticiser. An alternative is to carry out accelerated ageing tests on the adhesives you consider using.

D Accelerated ageing

One way of estimating the ageing characteristics of an adhesive is to carry out processes which accelerate the effects of time, one of which is to observe what happens at a higher temperature. Such programmes of materials testing are quite often used (and, when carefully controlled, they can give strong indications of the likely natural behaviour). Unfortunately, the approach is fraught with problems because a variety of factors determine the ageing behaviour of a joint and because by their very nature, the process is artificial. As you will appreciate by now, the important processes of degradation involve either chemical reactions of some sort, or physical changes that depend on diffusion. Usually there are a number going on simultaneously.

One important and early consideration for accelerated ageing that involves heat is the T_g of the materials being aged. Obviously, the materials must not be melting at the temperature selected for accelerated ageing. In practice, any heat ageing must be carried out at a temperature 10°C below the T_g or the lower end of the T_g range. For many acrylic polymers used in conservation, this puts an upper limit on the ageing temperature of 30–35°C. This, however, is the indoor temperature of a museum building with a degree of passive climate control, during a hot summer, and it can be reached in a gallery with roof lighting even in northern Europe. For a tropical climate and an uncontrolled building, it may not even be the summer temperature range. It is normal service conditions, not accelerated ageing, in effect.

You know from Section A in this chapter that the rate of diffusion increases very rapidly with increasing temperature, and chemical reactions exhibit the same. The rate of diffusion and rate of chemical reaction are controlled by the activation energies of the processes. As you learnt in *An Introduction to Materials and Chemistry*, only a small number of molecules have sufficient energy to cause a chemical change on collision with another molecule. Similarly, only a small number of

Table 5.4 Illustration of reaction rates at elevated temperatures, for chemical reactions that proceed at ×2 and ×3 their rate at 20°C.

Temperature (°C)	*Rate of reaction A, compared to rate at 20°C*	*Rate of reaction B, compared to rate at 20°C*
20	1	1
30	2	3
40	4	9
50	8	27
60	16	81
70	32	243
80	64	729

molecules will have sufficient energy to squeeze through a gap in a polymer network. As the temperature rises, the proportion of sufficiently energetic molecules rises dramatically (exponentially, in fact). The dramatic effect of this is tabulated in Table 5.4 using two hypothetical chemical reactions that proceed at the same rate at 20°C, but have different activation energies.

The rate of reaction A is assumed to increase by a factor of two for each 10°C rise in temperature, and that of reaction B by a factor of three (both are reasonable assumptions but will not be universally true). Table 5.4 illustrates both the usefulness and the main disadvantage of accelerated ageing programmes. First, such a process carried out on a material in which, say, reaction A is occurring will, in principle, provide the result of about five years' worth ageing at 20°C in only a month at 80°C. Unfortunately, in most materials several different thermally activated processes occur simultaneously at normal temperatures. Since they all have different activation energies, accelerated ageing will exaggerate some reactions more than others. You can see from Table 5.4 that reaction B, which has the same rate as reaction A at 20°C, is proceeding more than 11 times faster at 80°C. If reaction B is the most important one in natural ageing and provided nothing else would happen to the samples at 80°C like slumping, melting, or rapid loss of solvent as plasticiser, the accelerated ageing will be useful, but if not, very misleading results would be obtained. Clearly, the lower the temperature at which accelerated ageing is carried out, the more reliable the results will be because the difference between the rates of the different reactions will be lessened. On the other hand, the higher the temperature, the shorter the time in which indications of unsuitability of some adhesives become clear. But in the worst scenario, after accelerated ageing at a higher temperature, *none* of the samples may look suitable for the task in hand!

Accelerated ageing to simulate discoloration or degradation in light which uses very intense light for a short period of continuous exposure to simulate lower light levels over a longer period also has disadvantages. Photo-oxidation of polymers is a stepwise series of reactions, some of which are slower than others. If a reaction occurs as a series of steps, the rate of the overall reaction is determined by the slowest step. This 'bottleneck' is correctly called the **rate-determining reaction**. Some of the stages in photo-oxidation are activated by light while some are thermally activated. Increasing the intensity of light may increase the rate of one step but this may not be the one that fully determines the rate of the complete reaction. It's also true that service conditions matter: accelerated light ageing of conservation materials is often carried out using UV-filtered lights, on the often reasonable grounds that museums filter out UV light anyway. The results of such ageing do not then apply to adhesives or coatings that will be used in some historic buildings that do not have UV filtering fitted, in the unfiltered and dramatically daylit entrances of some modern museum buildings, and certainly not for outdoor sculpture exposed to both UV light and weathering.

It's necessary to bear in mind that the service conditions of adhesives may include exposure to atmospheric pollutants, usually indoor pollutants but in some cases the outdoors ones too, in a building with open windows, or for an outdoor sculpture. Accelerated ageing with pollutant gases is still not very common outside a university context since it requires specialist equipment to contain and monitor concentrations of gases, some of which are toxic. That makes it no easy matter to find out whether an adhesive has much sensitivity to pollutant gases. At the time of writing, this is an area that needs more research before even general guidelines can be given on its importance.

Bearing these reservations in mind, accelerated ageing can give useful pointers to those factors that are important in degradation. The process is likely to be more effective if the adhesive is tested in conjunction with the material with which the joint is to be made, since diffusion of chemicals into the material may influence degradation processes. Ageing a realistic mock-up at least gives the opportunity for the chemical reaction or the process you have not considered to have an influence on the results!

Measuring or monitoring some of the other characteristics of an adhesive is possible. In particular, the shrinkage of an adhesive can be examined by pouring it into a wooden frame (or a frame made from paper straws if a delicate frame is required to mimic the job in hand) so that an attached film is formed. The subsequent distortion can be easily observed. Equally, the chemical reactivity between a material and an adhesive can be investigated by putting them in close proximity in, say, a test-tube. Such trials give indications about which adhesives and materials are chemically incompatible, but again this is often carried out at an elevated temperature.

You will now see that a variety of testing scenarios can be devised for individual jobs; many of them are useful although it is important to bear in mind that they are unlikely to afford accurate predictions, only indications. Consequently it is unwise for a conservator to become too absorbed in carrying out detailed work of this kind; their relevance is limited and their interpretation difficult. Time might be better spent on a careful reading of the manufacturer's information on the properties of the adhesive or coating, or in discussion with the technical advisor working for the manufacturer (rather than the distributor of the product, who will have less knowledge), to find out if your application is similar at a process level to the application of one of their regular customers.

And finally, nothing beats natural ageing! If you are making test samples or mock-ups of any kind prior to treating an object, it is very useful to retain a set in the long term. These should be kept in broadly similar conditions to the treated object: this might be exposure to UV-filtered light but no sunlight in a corner of your studio, or in a wooden cabinet just like the object's storage, or in a tightly controlled environment in the museum store, but placed near the air extraction grille to give both dust deposition and exposure to internal pollutants. Many students and emerging conservators have found such samples very useful in their mid-career research.

6 Coatings

An amazing number and variety of coatings have been used on objects over the centuries, and all sorts of additional terms have been employed to describe them (paint, emulsion, varnish, lacquer, enamel, glaze). A coating is a thin film – just like an adhesive. This chapter is concerned mostly with coatings applied by conservators for preservation purposes, and these are usually not coloured. It is not discussing paints or decorative finishes in any detail.

A **paint** is a suspension of solid pigment in a suitable liquid medium which when applied to a surface dries to form a solid film. This makes it sound like the retouching medium used by many conservators; however commercial paints are usually more complex in formulation and more strongly targeted at both a particular application and a specific substrate to be painted. Even historic, hand-ground paints are often more complicated in formulation than this two-component description. A commercial paint always includes additives to improve its handling or **rheological properties**. An **emulsion paint** is one in which the paint itself (the mixture of polymer and pigment) is in the form of finely dispersed particles in suspension in a liquid. (Emulsion coatings are pretty rare, because they are not needed.) **Varnishes** are coatings which do not include coloured or opaque white pigments, but which may have a dye dissolved in, to give a coloured coating. Earlier in the twentieth century, some types of coatings were often called by other names, such as 'enamel paint' or 'enamel' to describe a paint suitable for outdoor use, or 'lacquer' for a transparent coating that would withstand weathering outdoors. Earlier twentieth-century outdoor paint products with 'lacquer' marked on the tin were usually glossy, not matt. In Southeast Asia traditional varnishes and lacquers were based on the sap from trees and have a very different chemistry and properties from modern conservation materials or these paints. Traditionally in the West, **enamels** and **glazes** were vitrified (glass-like) coatings produced by a firing process, not applied coatings at all. These are all good reasons to avoid these somewhat date-specific and culture-specific terms. Coating is the most general term and the most useful.

A The requirements of a good coating

This chapter is short because the requirements of a coating and the means by which it is produced have a very great deal in common with those for an adhesive.

DOI: 10.4324/9781003389217-6

For the main part, you have, therefore, already covered the necessary science and its applications as well as the usual materials that are used for conservation coatings (but not for modern paints). Consequently all that needs to be done in this chapter is to pull together those aspects that are important to coatings and to discuss a few topics that are peculiar to coatings – gloss and matt appearance, for instance.

A useful way of getting to grips with the requirements of coatings is to list the ones that are common to the properties of a useful adhesive. As you did in Chapter 1, do this under the headings of the requirements of the completed coating and the requirements of the coating process.

A1 The completed coating

In fact all the requirements of an adhesive are relevant to a conservation coating.

- *Is it strong enough?* A coating must adhere to the surface of an object, at least so that it is held firmly in place. The adhesive bonding may also need to withstand internal stresses created by any differential thermal expansion between the coating and the underlying material, or by shrinkage. If these are present, the coating may crack, cup, and later flake off. A coating is often required to be hard – to withstand abrasive wear or to avoid dirt being retained – and to be tough. All of these requirements are similar to those already considered for adhesives in Chapter 4.
- *Should it be flexible like the substrate it is applied to?* The short answer is, yes.
- *Can it wet and cover the whole surface of the object, with no thin patches or uncoated gaps?* This is a difficult thing to ensure in practice. The application of several thin coatings to achieve a fully covering one is worth considering.
- *Should it be invisible or obvious?* For coatings used in conservation the answer is usually straightforwardly that it should be transparent, in fact as invisible as possible, in order fully to reveal the object's surface.
- *Will it harm the object?* The same arguments apply to coatings and adhesives. Objects can be damaged by the release of chemicals from the coating, for example, nitric acid from cellulose nitrate, or by the effects of heat. Molten materials are seldom used as coatings, except perhaps waxes, but coatings that are formed *in situ* by chemical reactions can evolve heat. This heat could be damaging, particularly by accelerating the oxidation processes that discolour pigments, dyes, and paper.
- *Will it last?* The requirements are similar to those for an adhesive (Chapter 5). The polymeric material is thin and spread over a large area which means that it is very exposed to attack by oxidation, light, and atmospheric pollutants. Many coatings are 5–20 μm thick when dry, and a coating 50 μm thick would be thick indeed. Since the object must be viewed through the coating, the coating must not discolour, nor must it either reflect or scatter too much light.
- *Can it be removed later without damage to the object?* This is a particularly important question for more fragile objects based on organic materials, such as

paintings, textiles, paper, and wood. As with adhesives, the answer is determined by the solubility of the aged coating or its ability to swell and consequently weaken, compared to the object's propensity to react to the same solvents.

A2 The coating process

- *For a good result.* The surface to be coated should be clean. This, as with adhesives, is necessary for good wetting, and particularly so for coatings, because dirt or fingerprints on the surface will detract from the appearance of the object. These will show more after coating than they did before.
- *The applied coating should spread easily and evenly.* To give the object proper protection, a coating should spread over the whole surface and have a uniform thickness. That means it must effectively wet the surface. It is applied more effectively as a liquid with a low surface tension and viscosity, as this avoids brush marks, which require time to level out. **Levelling**, the ability to form a smooth surface before the tacky stage, is nearly always desirable in a coating, but much less so in a paint, whose brush marks may be utilised expressively. Levelling may not occur before tackiness if the room or site temperature (or even the substrate temperature in the case of a metal sculpture in the sun) is a lot higher than typical room temperature.
- *Setting time.* Again, as with an adhesive, the final product has to set. If it remains tacky for a long time, there is the possibility of dust being entrapped and marring the appearance of the coating.

A3 Optical properties

As you have seen, many of the basic requirements of a coating are common to those for an adhesive, and, therefore, many of the arguments that have been used in Chapters 3–5 are valid for coatings. However, some coatings need to have certain characteristics not associated with an adhesive. In particular, in conservation work coatings are usually used to protect an object in such a way that its features can be displayed while still providing adequate protection. Consequently its **optical properties** are important; specifically, it should be transparent, and decisions have to be made about whether it should have a gloss or a matt surface finish, and how noticeable (or not) it should be. In addition, since a coating is usually extremely thin, its effectiveness as a barrier film is often of considerable importance. These characteristics will be considered in turn. The very important topic of the **refractive index** of a coating was mentioned earlier.

A surface can be seen as a result of the light that is reflected off its surface. A transparent coating allows most of the light passing through it to be reflected back by the surface of the object beneath it, and back through the coating, while a small amount is reflected off the surface of the coating itself. Figure 6.1(a) illustrates this for a flat object and a glossy coating. This is the 'ideal' case. Colourless polymeric materials applied as a thin film to form a coating can appear to us as anything from highly transparent to completely opaque. Any flaw, such as an ageing crack or a loss due to flaking, inhomogeneity within a coating material caused by air bubbles, or a change in its thickness (obvious brush marks give a 'bumpy' surface to the coating) **scatters** the light falling on it in a variety of directions, so the viewer sees less of the object through the coating, as in Figure 6.1(b): the coating looks less transparent than in the ideal case. In very flawed and bubble-filled coatings, not much light actually gets through to the surface of the object to be reflected back, and the coating is the only part of the object that the viewer really notices. The light scattered at a range of angles looks more white than object-coloured.

A flawless, colourless, amorphous polymer is transparent because there are no internal discontinuities to scatter the light passing through it. On the other hand, crystalline polymers are normally very slightly translucent because of the mixture of crystalline and amorphous regions they contain. This tends to mean that if transparency were the only consideration, an homogeneous coating of an amorphous polymer is required. But this is a more subtle effect than the effect of the coating's surface on a micro-scale and is unlikely to influence the choice of polymer for the coating in practice.

Gloss is a term for describing the optical nature of the surface (and only the immediate surface) of a material, whether it is transparent or opaque. A very **glossy** surface is one that behaves like a mirror; it has a very flat surface, and so the angle of reflection matches the angle of entry of the light. In contrast a **matt** (or **matte**) surface is one that is rough on the micro-scale, so that light is scattered in all directions by the surface irregularities, as in Figure 6.1(c). In other words, gloss is neither colour-dependent nor material-dependent. It can be measured with a glossmeter, which is discussed in more detail in *Colour, Colour Measurement and Colour Change*. Gloss measurements are more perceptual and comparative than colour measurements, in that it is not possible to form a mental picture of what a gloss reading must 'look like' on a given surface.

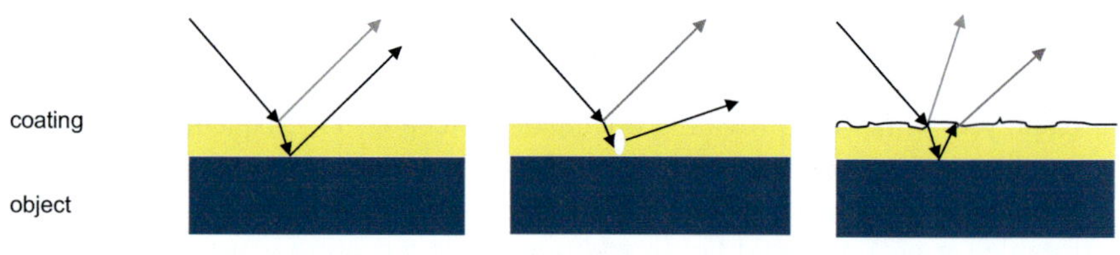

Figure 6.1 (a) and (b). Schematic of a transparent coating (pale yellow), with a flat and glossy surface, on an opaque object (blue). (a) Some light (pale grey) is reflected off the coating at the same angle, but most goes into the coating, its direction of travel refracted by the coating's refractive index, which is greater than the RI of air. It is reflected off the object's surface, and emerges parallel to the (grey) scattered light. (b) Some light (pale grey) is reflected off the coating as before. A bubble in the coating scatters light out of the coating at a non-parallel angle. The object won't be as easy to see through the coating. (c) The coating is matt, with a rough surface. Some light is scattered off the surface (pale grey) and the angle of scatter varies from point to point on the coating. The light reflected off the object itself comes out of the rough surface of the coating at varied angles too. The viewer is going to see less light reflected, making the object harder to see.

To obtain the mirror-like surface required for a glossy finish, the molecules in the coating must be sufficiently mobile and have sufficient time to move so that the surface can fully level before the thin film that constitutes the coating has dried or set. This means that the coating must remain above its T_g for long enough to allow polymer movement, while the solvent evaporates slowly. The reverse requirements produce a matt finish; that is, a quickly evaporating solvent allows the T_g of the coating to be increased very quickly to above room temperature so that movement is restricted. However, matteness can arise either from regular surface texture (most woven fabrics look matt, lustrous ones being produced only from particular thread types and weave structures), or from internal haze, cloudiness, or 'non-transparency'. Producing a regular surface texture in a coating is challenging and not very practical to attempt outside a controlled industrial environment.

Instead, it is easier to include **matting agents** in a coating, to provide the haze. These work like extenders do in a paint formulation. The matting agent should have a fairly close refractive index to the coating itself, so that each particle scatters a little light, in varying directions. The coating looks a bit hazy compared to one without any matting agent, and adding more matting agent to the liquid coating will increase the hazy effect. Sometimes particles are used, such as chalk or mica, or micronised silica. Glass micro-balloons would work optically too. The disadvantage is that all of these additions tend to affect the rheology of the coating, and most obviously its levelling and spreadability. Sometimes in the past another, but non-compatible, transparent polymer was added, often melted wax. In some situations, this would compromise the ability of the coating to wet the object's surface, which scatters even more light, but reduces the protective function of the coating, as well as altering its solubility when the coating needs to be removed by a later generation of conservators. The particle route is usually a better one to take.

A4 Protection

A protective coating can act as a barrier in two ways: to keep an object clean and free of dust that might eventually form a layer that is challenging to remove safely without scratching the object's delicate surface, and to protect it from attack by chemicals in the environment. The former function requires a relatively hard coating that, when necessary in future decades, can be removed readily. The protection aspect might be against water, atmospheric pollutants such as hydrogen sulphide (H_2S) or sulphur dioxide (SO_2) that would attack some metals, or oxygen (or even the urban pollutant ozone) that could react with an object made of organic materials. In order to attack the underlying material, these gases must diffuse through the coating. Ideally, a coating should not be permeable but, as you know from Chapter 5, this is impossible for small molecules in polymeric materials. The problem is made worse by the fact that coatings are so thin. So to provide the best barrier to diffusion, the molecules in a coating should be packed as close together as possible. In this case, therefore, a crystalline polymer is preferable to an amorphous polymer, and it should be below its T_g, in relation to the storage or display temperature for the object. And if the coating is thin in patches or not applied as a completely covering film, then these materials have access to the surface of the object.

As with adhesives, coating materials have to start out in liquid form and end up as a solid. Again, the various types can be classified as melt-freeze, those that depend on drying by evaporation and those that are cured by chemical reaction(s). The first type is rarely used and is limited to low melting point waxes. The waxes are soft when applied, and to get a gloss finish the molecules must be moved and arranged mechanically to produce a mirror-like surface by polishing.

To illustrate some other important requirements of a coating, and also to continue the application of the polymer chemistry you learnt in Chapter 2, consider a few examples of each of the other two classes of coating materials.

B Solvent coatings

In Western Europe, **shellac** dissolved in alcohol is one of the older coating materials. It's even been used as an adhesive for broken ceramics, although it could never provide an 'invisible' repair since it is pale golden at best, and brown when aged in light. It is a resin produced by an insect, the lac insect, to protect its eggs. Shellac is a complex mixture of thermoplastic polymers which with ageing develops cross-links and becomes progressively less soluble; thus it becomes increasingly difficult to remove by dissolving it.

Cellulose nitrate was the first important synthetic 'lacquer' and is still in use in some museums as a coating for objects made of silver, to slow the rate of tarnishing. (Because it was used for automotive finishes in the earlier twentieth century, and they had to withstand the weather, it was described then as a 'lacquer', and the name has stuck in manufacturers' product literature.) Its structure and advantages and disadvantages as an adhesive were described in Chapter 3, Section B. This material shrinks through loss of solvent and plasticiser and thus becomes hard and brittle. It also degrades chemically by releasing small amounts of nitric acid. It is still used, primarily because it is easy to remove when it has yellowed and acquired a patchy appearance. Widely used cellulose nitrate products included Frigilene® and Ercalene®.

Better chemical and mechanical stability is provided by **acrylic resins** which are based on the methacrylate family of polymers. The best-known example today is Paraloid® B-72, used for decades as a conservation picture varnish. It remains soluble in any of the several solvents that can be used to apply it, it is easily replaced, and compared to many other acrylics polymers used for coatings it is hardly affected at all by UV-free light. However, its solvents are all quite polar, making this polymer usually unsuitable for solvent-sensitive objects such as some plastics. For some tasks it is useful to be able to choose an acrylic resin that is soluble in solvents that are not so powerful as to attack the coated material. The long side-chains in poly(lauryl methacrylate) allow it to dissolve in white spirit, which is a relatively innocuous solvent from the point of view of many classes of organic materials. (For some conservators however, white spirit is an allergic sensitiser.) Other methacrylates in which the methyl ($-CH_3$) side-group on the oxygen atom is replaced by a large hydrocarbon side-group, for instance, a butyl group ($-C_4H_9$), as found in the once popular Paraloid® B-67 (which yellows significantly more than B-72 and cross-links over time, becoming more insoluble in the solvent used to apply it), or even a lauryl group ($-C_{12}H_{25}$), might be more suitable here. These large side-chains give a high degree of mobility to the molecule because they are themselves flexible, and so they keep the backbone chains of neighbouring molecules further apart. They also make the polymer more soluble in hydrocarbon solvents, since they have a very similar structure and composition to these side-chains (remember that 'like dissolves like'). As polymers can be made with various proportions of these units fixed onto the main chain, there are enormous variations possible in their composition and properties. These polymer solutions have potentially very satisfactory properties if their lower lifetime before yellowing is acknowledged, and they are widely used in commercial coatings. (Commercial coatings are never intended to last as long as materials developed for the needs of the conservation profession: their manufacturers, after all, hope for repeat sales.)

A point to be careful about when using a solvent-based coating on some objects is that it may not fully and properly coat the surface it is meant to protect. After any liquid film has been spread over a rough solid surface, it will not remain of uniform thickness. This is due to surface tension, which as you know, pulls a liquid surface into a curved meniscus. This means that the film becomes thinner on the peaks and edges of a surface. Thus, the areas on the surface most susceptible to damage and chemical attack are barely protected by a coating based on a liquid system. Emulsions are more successful in this respect than ordinary solvent coatings. The poly(vinyl acetate) (PVAc) co-polymers that were described in Chapter 3, Section B, are one of a number of coatings used in the form of an emulsion. Emulsions have the added advantage that the carrier is usually water, which means that they avoid the toxicity and flammability hazards of volatile organic liquids. Another way of tackling the problem is to use liquids with a higher viscosity that react chemically to provide the solid coating. 'Classical' emulsions, however, tend to have a hazy appearance.

C Reaction coatings

Like reaction adhesives, reaction coatings involve polymerisation *in situ*. There are two basic types, those in which network polymerisation occurs as they dry out, by means of oxygen in the atmosphere, and those in which the polymerisation occurs by chemical reactions between polymers in the coating material.

Traditional **oil-based** paints and twentieth-century oil-based house paints are the most important example of the first type. The latter used natural fatty oils such as linseed, soya, and tung, each of which contains a mixture of different long-chain molecules. (They are actually esters formed between glycerol and various fatty acids. Esters were introduced in *An Introduction to Materials and Chemistry*.) These molecules, if straightened out, would be roughly Y-shaped, each arm of the Y containing 18 or more carbon atoms with one, two or three carbon–carbon double bonds in the chain. As these oils dry out, oxygen diffuses in and reacts with the double bonds to form cross-links. An oxygen atom links two molecules together by utilising one of the bonds in the double bonds. It is very similar to the way sulphur atoms cross-link rubber molecules in the vulcanisation of rubber (as shown earlier in Figure 2.10). In this way the oil is converted into a network polymer which, as more links are formed, becomes progressively more rigid. However, the process is slow, and drying of the surface which has the best access to oxygen slows down the diffusion of oxygen to the body of the film to continue the process, which means that the paint can remain tacky for days, making it liable to pick up dust. With time, the extensive cross-linking which occurs leads to hardness and brittleness. In addition, oxidation reactions with impurities produce molecules with a strong yellow colour. Both of these changes make films of oils like linseed and tung unsatisfactory for contemporary use as a decorative finish, and other polymers are now used in house paints. Chemical adjustments can be made to the ingredients in these paint formulations to achieve quicker drying and to reduce the amount of oil used, and hence to reduce some of the yellowing. Many such formulations incorporate polymers called **alkyds** which are a form of polyester (see Chapter 2, Section C).

The other type of reaction coating, involving network polymerisation between components in the coating mixture, is similar to reaction adhesives. The kinds most commonly used on objects in the past were based on epoxy resins, polyurethane, and urea formaldehyde. As was discussed in the chapters on adhesives, the difficulty of removing them presents such a major issue for future object treatment that they are not recommended as coatings today.

7 Consolidants

Consolidation is widely used in conservation as a means of imparting structural strength to an object that has deteriorated and is in danger of disintegrating; crumbling stonework, embrittled and aged canvasses for painting supports and embrittled leather book bindings, and worm-eaten wood are examples. In a material requiring consolidation, many of the primary and/or secondary bonds that hold it together have been broken, to the point where the structure can no longer support its own weight (or any other forces it has to resist) and retain its shape. For these reasons worm-eaten wood becomes soft and spongy to the touch and deteriorated stone crumbles away to a powder. Consolidation is an artificial way of repairing the damage caused by what are, after all, natural processes.

The first requirements of a successful consolidating material (a **consolidant**) are that it can impregnate the object, at least to the depth of the weakened outer area but preferably into the non-deteriorated material beyond, and impart strength by binding everything together – in other words, it must act like an adhesive. In fact, thinking about consolidants is very much like thinking about adhesives in three dimensions. But it should not act like a coating. It is not meant to be visible at the surface. To avoid going through all the questions previously asked of adhesives and coatings, you could simply ask which of the requirements for an adhesive joint, and the joining process, are *not* requirements for a consolidant or the process of applying a consolidant.

By using Figure 1.2 in Chapter 1 as a checklist, you should have found that the answer is simple: all the requirements for an adhesive and for the jointing process are relevant to consolidation. However, a bit more attention needs to be given to the consolidation process because the requirements are greater than for adhesives. In joining or coating, care has to be taken to ensure that the conservation material covers the whole of the portions of the object that require treatment. This, of course, applies in consolidation, but achieving it is rather more difficult.

Clearly, for reasons identical to those for adhesives and coatings, a consolidant has to start out in liquid form and end up as a solid. However, the liquid may have to penetrate into very long, and very tiny, crevices and pores in an object, and this makes the task of a consolidant more demanding.

It is necessary for a consolidant to wet the surface of the material on which it is being used, not only to make a satisfactory adhesive bond, but also to enable it

DOI: 10.4324/9781003389217-7

to penetrate the object. But at the same time, it has to wet both pristine material towards the core of the object and deteriorated and possibly chemically altered material at the surface. If you visualise a porous solid as a bundle of capillary tubes, you can appreciate that a liquid consolidant will be drawn into a porous material by means of capillary action. Although the surface tension determines the extent to which a consolidating liquid penetrates a porous solid, other factors affect the rate of penetration. The two main ones are the viscosity of the liquid and the size of the pores.

A Consolidation and its purposes

The reason that excavated material requires consolidation is that its environment changes suddenly and rapidly as soon as an object is excavated. Many archaeological investigations dig below the water table, meaning that objects are recovered waterlogged from saturated soil and suddenly find themselves surrounded by air instead, usually with a lower moisture content. Once objects are found and excavated, the ultimate aim is to store them in a museum or repository that will have more typical and museum-like environmental conditions. From the perspective of the newly discovered object, it is moving to much drier conditions, and treatment must enable it to dry out at a suitably slow rate, losing a lot of moisture without deformation. Or conversely, an object moved out of an enclosed burial chamber in a hot, arid climate is going to move to air with much higher relative humidity as it is brought out, and then must continue to exist with an increasing moisture content until it acclimatises to its future storage or display conditions. Classical consolidation treatments were concerned with mitigating the deformation and separation of different materials within the object caused by this sudden shock followed by the drying/re-hydrating process. And since most materials of archaeological age have severely deteriorated and have irreversibly changed in appearance over centuries or millennia, a consolidation treatment is 'only' aiming to stabilise them for a long life in a museum-type environment and not to restore the appearance they had when first made.

Building conservation is a little more complicated as regards consolidation, but even here, consolidation may well be the secondary aim of a conservation intervention. A building requires structural strength to continue its existence, and any severely weakened building blocks or other components are going to be replaced with new material, in order to preserve the whole structure for a long period, and perhaps against stresses caused by climate change, for which it was not designed. Consolidation treatment will then be carried out on the less-damaged components *in situ*, and the aims of restoring some lost structural strength and improving surface appearance may have roughly equal importance. The preservation and treatment of interior decoration, plasterwork, flooring materials, and the like are more likely to involve consolidation as the major conservation procedure, after the structural work is complete and long-term survival is thus assured. There is always the option of removing a small surviving piece of decorated plaster or wood for display and interpretation of its construction and likely original appearance. In effect, the materials being treated morph into a museum object that need not fulfil its original structural function any longer.

This author, updating the present text, felt that the original text was almost unconsciously written with the aforementioned scenarios in mind.

But the treatment being planned may in fact be more like the consolidation of flaking surface decoration to the preparatory layer that was applied to a support. This might be required because inherent vice caused the flaking in the first place (the artist or maker did not use a compatible preparatory layer and paints, or took no account of drying time for the different layers, or applied some layers very thickly and hence caused uneven drying); or because fluctuating environmental conditions have eventually led to cracking and material failure; or because of a sudden disaster such as immersion in floodwater. In these cases, like the removed painted plasterwork, aesthetic considerations determine the success of the consolidation treatment. A good intervention should not be noticeable at all at normal viewing distance: the visitor or curator should not be painfully aware that there had ever been any damage, at first glance. This type of 'rescue' treatment is still challenging, because the object being treated is already weakened and its mechanical properties may now be distinct from what would be expected to have changed due to more typical ageing processes.

The greatest consolidation challenge is to treat a very delicate and friable, underbound, matt, painted surface on a support that is in itself both delicate and highly responsive to environmental fluctuations and liquid water. Painted barkcloth would be an example of this challenging class of material. There are several conflicting demands on the ideal consolidant for this task: it must penetrate into the support to provide attachment of the paint and cohesive strength, while matching essential properties of the support such as flexibility and dimensional response to moisture; it must adhere crumbs of thin paint one to another; and it must provide a thin coating to hold everything together – but without affecting the intended matt appearance in the slightest. Choosing the consolidant is a challenge to be sure, but devising the optimum application method in the circumstances can be an even harder task and more critical to the success of the treatment.

It's no surprise that late twentieth-century conservation literature offered many examples of the successful consolidation of archaeological objects with conservation materials that are falling out of use due to health and safety concerns, whereas more contemporary conservation literature offers discussions and case studies on the consolidation of challenging, friable surfaces, with more emphasis on application method and visual/instrumental assessment of the post-treatment appearance than on the actual consolidant used, and the choice of *that* one over another one. There is acknowledgement that hand skills and acquired experience in application by any method are the key skills that conservators bring to such interventions, and that sometimes there are several materials that would be suitable for the task, provided they are in the hands of a skilled user who is sympathetic to the material being treated. Finding out materials with the right properties has become easier through online searches of manufacturers' websites, publications, information on the chemistry of materials novel to conservation, and so on. It helps to understand the chemistry and physics behind consolidation, that extra-complicated case of adhesion and the application of coatings, and this book has aimed to provide tools to enable such understanding – but that can only be the beginning of the conservator's work.

The following sections discuss the properties of consolidants that matter in terms of choice of consolidant type. The concluding section of this chapter – and of the book – looks forward to likely developments in the materials available for consolidation, and hence for adhesives and coatings too.

B Viscosity

As you know from Chapter 1, Section E, the **viscosity** of a liquid is a measure of its mobility (runniness, ability to flow freely); water is more mobile (less viscous) than a water-based house paint, for example. When a liquid flows, its molecules slide over one another. This is a process of shear, and viscosity is, essentially, a measure of the rate at which shear strain occurs under a shear stress. In the absence of other sources, such as an applied pressure or capillary action into pores, the shear stress is provided by the force of gravity – which is why rivers run down to the sea, water flows over a waterfall, and you can pour milk from a jug. Clearly the stronger the bonds between the atoms or molecules in a liquid, the smaller will be the shear strain produced by a given shear stress; in other words, the more viscous it will be.

Table 7.1 lists the viscosities of a wide range of liquids at room temperature. All of the specific materials in this table have been used by restorers in the past, in some country and in some culture, but not all are used today by conservators, in order to protect both their own health and that of the object. They are listed here to demonstrate a wide range of viscosities and to indicate why these materials were used in the past despite shortcomings or hazards that seem obvious to us today. The

Table 7.1 Viscosity of a range of materials once used by restorers and/or more recently by conservators.

Material	*Viscosity at 20°C* ($Nsm^{-2} \times 10^{-3}$)
amorphous polymer (T_g = 20°C)	10^{15}
bitumen (as a solid)	10^{11}
amorphous polymer (T_g = 0°C)	10^{10}
glycerine	1500
castor oil	1000
linseed oil	30
epoxy resin (before it sets)	15–25
ethanol	1.2
water	1.0
triethoxymethyl silane (stone consolidant)	0.6
trimethoxymethylsilane	0.5
acetone	0.3
diethyl ether	0.2

units have been chosen for easy comparison with the best-known liquid, water, whose viscosity has a value of 1.0. Notice the enormously wide range between, say, bitumen[1] and ether. Most organic solvents are more free-flowing than water in fact, but that is a surface tension effect created by the strong hydrogen bonding in water: water has unusual properties, as was discussed in *An Introduction to Materials and Chemistry*.

The viscosity of all liquids decreases as the temperature is raised; they become more free-flowing. This is why, for example, engine oil flows more freely when the engine is warm. This effect is another illustration of the effect of heat on a process. With increasing temperature a material gains more thermal energy. This means that there is an increasing probability that thermal activation will overcome the bonding between molecules. You could carry out for yourself a demonstration showing the influence of viscosity on the rate at which a liquid penetrates a crack or pore.

Demonstration

Fill two beakers with a fairly viscous liquid such as glycerine, and heat one of them to about 80°C on a hotplate, handing it with due care and wearing eye protection. Place an identical capillary tube in each beaker and observe what happens. The hot glycerine will rise up the tube very quickly; the cool glycerine rises very slowly but will eventually reach the same height.

The *height* to which a liquid rises in a capillary is determined by the surface tension (the surface tension of a liquid decreases with increasing temperature, so the capillary rise for hot glycerine will be a bit less than that for the cool glycerine). The *rate* at which the liquid rises to that height is determined by the viscosity. Therefore, to achieve good penetration of an object by a consolidant, a liquid of *high surface tension* and *low viscosity* is desirable. The practical benefits of raising the temperature of the liquid to achieve lower viscosity in conservation treatments are dubious because it is often ethically inappropriate to heat an object, and also difficult in practice to heat it evenly through its thickness and to control temperature during heating. Usually a normally studio-based consolidation process on an outdoor site, whether in chilly winter or hot summer sun, would illustrate the effect of temperature on the whole procedure.

As a first step, consider viscosity by modelling it with a simple equation. Quantitatively, you can say that the amount of liquid passing into a porous medium in one second (call it Q), is *inversely proportional* to its viscosity, which is often called η (the Greek letter *eta*). That is:

$$Q \propto 1/\eta$$

This is read aloud as 'Q is proportional to the inverse of eta'. Here $\propto$ is a proportionality sign; it means Q is proportional to $1/\eta$ or, 'there is a direct relationship between Q and $1/\eta$'. As η gets bigger, $1/\eta$ gets smaller and so Q gets proportionally smaller. The formula simply describes what you would expect, that is that the rate of flow (Q) increases as viscosity (η) decreases; water flows more rapidly than oil, for instance.

The extent and rate at which a liquid penetrates a porous material depends on the characteristics of the material as well as those of the liquid, in particular on its *porosity*.

C Porosity

The porosity of a substance may be thought of as the percentage of pore space in the total volume – that is, the space not occupied by solid matter. It is possible that not all the pores are connected with one another or with the surface, so they may not be accessible to the liquid consolidant. The **effective porosity** is the pore space connecting directly or indirectly with the surface, and clearly for good consolidation the effective porosity should be as high as possible. With wood, for example, decayed organic matter often blocks the pores or narrows their width, thus making effective consolidation less easy. Figure 7.1 suggests porosity in stone: the pores vary in size and shape, and only some give access to the outer surface. This 2D sketch looks to have less than 20% porosity, but obviously porosity refers to the volume of the stone, not to a cross-section through it; you have to imagine that some of the pores here point into the page, and others out. Typical values for the porosity of sandstones (which are more porous than hard rocks such as granite) are

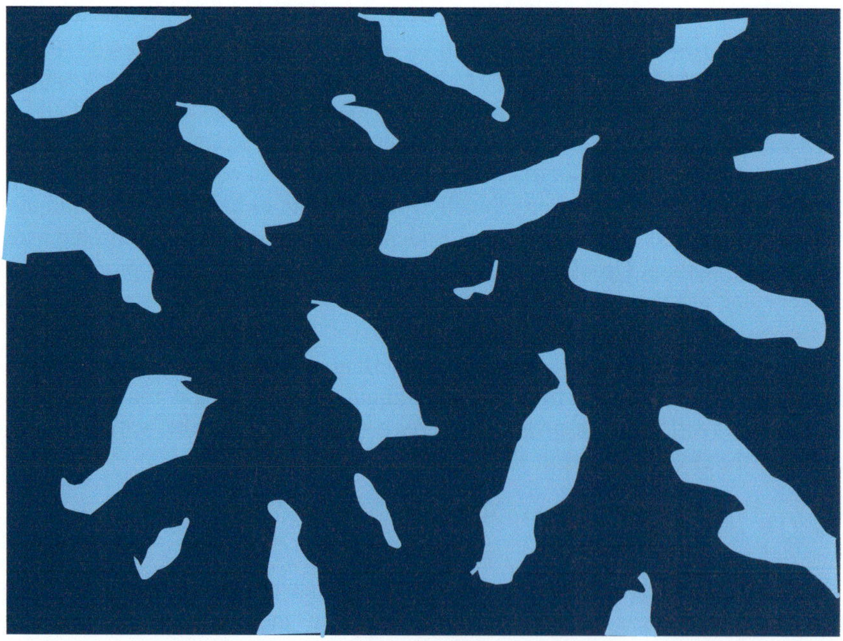

Figure 7.1 Schematic of pores (lighter blue) in stone.

15%–30% by volume and for limestones 5%–15% by volume. The extent and rate of penetration of a porous medium by a liquid also depends on whether the liquid is forced in under *pressure* and on the dimensions of the pores.

The rate of flow can be increased by exerting pressure on it. Pressure is similar to stress in that it is a force acting over an area, and it has the same units, Nm^{-2}. The effect of pressure is obvious; you can remove the liquid in a tube, such as a straw, much more quickly by blowing down the tube. You also have to suck harder on a straw dipped in a viscous milk shake than in a glass of lemonade. In each case you are effectively overcoming the viscosity of the liquid by exerting pressure.

For similar reasons, to raise a specified volume of liquid through a straw you would have to suck harder on a long straw or a thin one than on a short one or a wide one. So the *rate of flow* of a liquid also depends on the cross-sectional area and the length of a pore.

D Factors affecting penetration of the consolidant

Having established that the rate of flow of a liquid in a porous material increases with pressure and also that it increases with the cross-sectional area of a pore, you know that as the *viscosity* of the liquid increases or as the *length* of the pore increases, the rate of flow *decreases*. The degree to which the flow rate actually varies through changes in pore dimension, pressure, and viscosity of liquid is an intrinsic property of the porous medium and is called its **permeability**. If the permeability of a solid is zero then there can be no flow – the substance is said to be **impermeable**, or impervious to that liquid. Some of the factors that affect permeability were discussed in Chapter 5 when looking at the transport of vapour through a polymer.

Since the intrinsic permeability of the object you may want to consolidate cannot be changed, other factors will have to be varied. The dimensions of pores in the object are predetermined and so only viscosity and pressure remain as variables. Viscosity can be altered by choice of material, by dilution in a less viscous solvent and by temperature change. Seeking ways of changing the pressure has led to the adoption of various consolidation processes such as **vacuum impregnation** in its different forms; details of the techniques that have been used can be found in the conservation literature. Do not forget that the force of gravity can also be used to promote consolidation, since a liquid is able to percolate down into a porous body.

E Consolidation methods in terms of chemistry

A consolidant has to be applied as a liquid and, as you know by now, there are three ways of doing this: melt-freeze, evaporation of a solvent carrier, and chemical reaction between consolidant liquids.

The first method has very limited application for deep penetration because most materials that could be used, waxes for example, have high viscosities when molten, and, further, their viscosity increases as they cool – which they will do rapidly as they penetrate an object that is at room temperature rather than matching temperature. The high temperatures necessary to keep wax molten during impregnation might well damage the object.

Consolidation using solids in solution is the 'classical' and (at first thought) the easiest consolidation technique, and in most cases the solid is a polymeric material dissolved in water or an organic solvent. This is, of course, a convenient way of reducing the viscosity; the main drawback is that the consolidant tends to be drawn back to the surface of the object as the solvent evaporates. This effect is the result of capillary action. Figure 7.2(a) shows a pore which has been penetrated by a consolidant in solution. On ceasing the impregnation process, the solvent begins to evaporate from the pore entrance. Due to the capillary action, the column of solution rises up the pore, just as it would rise up a capillary tube (see Chapter 1, Section E), and it leaves behind a void that is not consolidated. Problems can also arise because the evaporation of solvent can cause the collapse of fragile materials, such as the cell walls in organic materials. Waterlogged wood is a case in point. The cell walls are so weak that when the water is removed they shrink and collapse under the resulting internal stresses unless they are supported by solid consolidant: during consolidation, the surface tension of the solvent can also rupture the cell walls. You can illustrate this happening for yourself. Even if the pores are not in danger of collapse there is a long-term preservation issue; the treatment has created a band of consolidated material extending in from the object's surface, with different properties from the unconsolidated portion. A sudden stress, such as an impact, might cause failure at the junction between consolidated and unconsolidated material and spalling off of the treated material.

This illustrates the conservator's dilemma: the success of a consolidation treatment cannot be observed directly in most cases. Attachment of a coloured paint chip to the off-white preparatory layer it flaked off from illustrates one of the few cases when successful consolidation may be seen in action. The consolidation of a large object (an extensive wall painting for example) or multiple objects (painted low-fired ceramic figures being excavated one by one from the same site) might well be preceded by extensive trails of different consolidants, solution strengths, and timing between applications, all on surrogate material or mock-ups that can

Figure 7.2 Schematic of (a) solvent-borne consolidant in pore immediately after treatment. It has not reached the far end of the pore. (b) The consolidant concentrated at mouth of pore after evaporation of the solvent.

then be broken or cut through and examined in cross-section with a microscope to assess and compare consolidant penetration. These practical experiments would lead to better decision-making for the actual treatment.

Demonstration

Tie a piece of thread loosely across the diameter of a small loop of metal wire, such as that used by children for blowing bubbles. Dip the loop into a soap solution to form a soap film on it. The thread will lie loosely in the soap film, Figure 7.3(a). Now burst the film on one side of the thread and observe what happens to the thread: this is the effect of a light puff of air that does not create a bubble strong enough to detach itself and float off. The soap film is pulled taut, as in Figure 7.3(b), by the surface tension acting on the thread, that is, it is under a (self-imposed) compressive stress.

The same effect occurs when the solvent evaporates from a piece of wood. The surface tension can be high enough to exceed the strength of the cell walls and therefore break them. For this reason the solvent used should have as low a surface tension as possible. In the later twentieth century, polyethylene glycol (PEG, Figure 7.4), a long-chain polymer, was found to be a good consolidant for

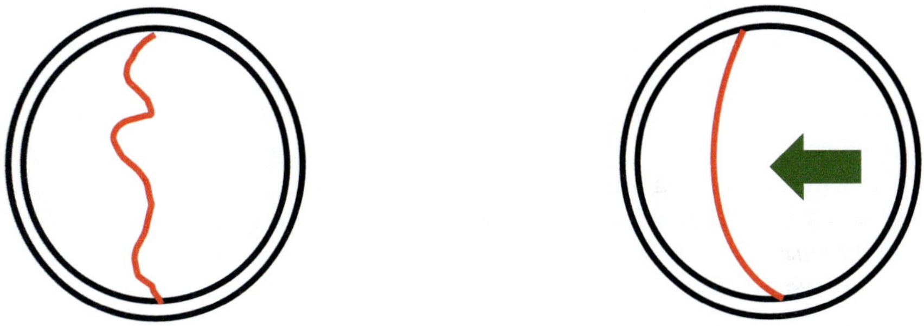

Figure 7.3 Schematic of (a) bubble-blower, ready for action. (b) Bubble forming due to effect of surface tension, but not yet free to float off.

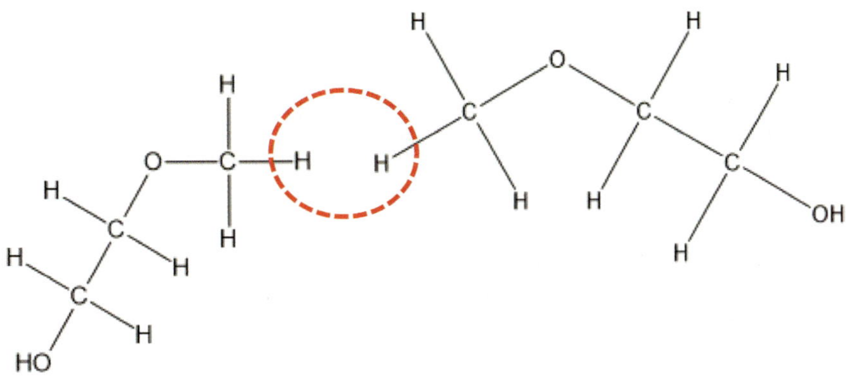

Figure 7.4 Structural formula for PEG.

previously waterlogged wood (once recovered from burial or even from mud or the sea) because it will readily form hydrogen bonds to the cellulose that makes up wood. With stone, however, it is less successful since there is no possibility of the formation of hydrogen bonds. There were many varieties in popular use, identified by their molecular mass; for example, PEG 1500 has a molecular mass of 1500. Although soluble in water it does not cause surface tension damage because consolidation has effectively taken place before evaporation occurs. And being used in water, it was and is an especially compatible treatment to follow washing in water, desalination in water (for objects recovered from the sea), and maintenance of high moisture content after excavation of waterlogged materials. (It has proved to have adverse reactions with some materials such as iron that may be found with wood, indeed reactions that compromise long-term survival of such composite materials, but that is beyond the scope of this book. And the fault lies with the nature of composite objects more than with the polymer in itself.)

As with reaction adhesives and coatings, consolidation can also be achieved by liquids that react to form a solid product *in situ*. Epoxy resins are an example. In the form in which they are generally available, the two components as produced by the manufacturer may be too viscous to achieve significant penetration, so they have to be dissolved in a suitable solvent as they are formulated into a commercial product. This can lead to a problem because some of the components of the consolidation mixture can be attracted more strongly to the material of the object than others; this will depend on the relative strength of the secondary bonding forces of the different consolidant molecules and those of the object. If the difference is significant, the components become separated and are not then all available in the required location to react to form the solid polymer. And it should always be borne in mind that use of a reaction consolidant like this is not reversible at all, even if a problem with the curing has emerged mid-treatment. Future conservators would in effect be dealing with a different, composite, material comprising the object, which would then have fewer treatment options than the conservator who chose epoxy resin had had before that treatment began.

This disadvantage is overcome if the components do not need dilution by solvent. One means of achieving this is by adjusting the chemical structure of the molecules involved so that a low viscosity is possible; epoxy resins are now commercially available with satisfactorily low viscosities. Indeed, most of those used as adhesives and gap-fillers intrinsically have low viscosities, and gap-filling as a process may require adding an inorganic material such as glass microballoons to the epoxy, to increase its viscosity to a more workable value for casting.

One way of ensuring a low viscosity is to have the consolidant initially in a form with the lowest possible molecular mass; that is, as the monomer from which the polymer network is created. A good example of this is the **organosilane** range of stone consolidants extensively used on building stone towards the end of the twentieth century. These simple organic compounds of silicon have a very low viscosity (see Table 7.1) so that deep penetration of the stone being treated is easily achieved, at least in principle. They have the ability to react with water to produce extensive network polymers: this is a virtue on site in many countries, where moisture is always present, given that the average outdoor relative humidity in temperate climates is far higher than in typical museum environmental conditions.

In the course of the polymerisation, reaction can take place either between a water molecule and a methoxy group ($-OCH_3$) with elimination of a methanol molecule and formation of a hydroxyl group (–OH), or between two newly formed hydroxyl groups with elimination of a water molecule. In this way water is produced and consumed. As a result, Si–O–C linkages are progressively converted into Si–O–Si linkages and a polymer network which has some similarity to the extremely chemically stable compound, silica, is built up. The extent of the polymer network will depend on the precise course which the reaction follows. As the reaction proceeds, the liquid monomer is converted to a solid polymer which shrinks back onto the pore walls and binds the stonework together. While this material may be used less often today, the concept of a consolidant which reacts chemically with the material being consolidated is a topic of current research in the 2020s, and treatment efficacy in terms of structural strength, water permeability, and the like of treated material is still compared to organosilane-treated material as a benchmark (Graziani *et al.* 2015; Briffa and Vella 2019).

The consolidant does not have to be a covalently bonded chain or network such as an acrylic or siloxane polymer. Stone can also be impregnated with a solution of calcium hydroxide in water. This slowly reacts with carbon dioxide from the atmosphere to form calcium carbonate, an ionic consolidant. (The very low viscosity of a gas, carbon dioxide in this instance, ensures that it can penetrate deep into the pores of the stone.) One danger of consolidation by means of chemical reaction is the tendency of some chemical reactions to liberate heat. If the quantity of heat evolved during consolidation is great, the safety of a friable object could be endangered.

F Consolidation methods in terms of application

'Classical' application of adhesives and coatings was by brush: a fine one to apply adhesives to breaks and a broader one for coatings. A fine brush would be used to drip-feed in more solvent adhesive to the adhered joint, to replace the shrinkage caused by solvent evaporation. 'Classical' consolidation relies on the same tools, often used first to adhere visibly detaching flakes with a fine brush or a drop of consolidant applied with even more precision using a syringe and a long, fine needle, followed by larger spots of consolidant to link together treated flakes, and then gentle brushing to provide coating and consolidant in one go. If the surface being consolidated has intrinsic, significant gloss, this method is ideal.

Difficulties begin when the surface to be consolidated is matt. Here, water-based consolidants which tend to have a lower refractive index on drying are often preferred. **Gelatine** and sometimes **sturgeon glue** have been traditionally used for this purpose: however the latter has supply problems and quality control issues and is not sustainable, meaning it is less popular today. The use of controllable local heating with a heated spatula or a custom-made small hot table is common in such treatments. Gelatine is often used for the local consolidation of flaking paint, even oil-based paint. It is particularly compatible with water-based paints that have a medium such as a plant or tree gum, or even water alone as the application vehicle.

Recently used materials for the consolidation of matt and/or friable surfaces have included ***Funori*** (Swider *et al.* 2005; Catenazzi 2017) and ***Junfunori*** (Geiger and Michel 2005). Both are derived from seaweed, for which there are concerns both about batch-to-batch variation, long-term ageing as a result, and the ethics of using a product that may not be sustainable if widely adopted. Other options are methyl cellulose adhesives such as the once popular **Methocel® C**; hydroxypropyl cellulose adhesives such as the **Klucel®** series; or more modern products such as the **Aquazol®** series (poly(2-ethyl-2-oxazoline)), in addition to traditional **wheat-starch adhesives** long used for treating paper-based objects. All of these can be applied in water, as can the Aquazol® series in ethanol or isopropanol.

Brush application of an aqueous consolidant is less easy given the surface tension of water, and some form of spraying, misting, or nebulisation might be preferable, such as the use of an ultrasonic nebuliser. A few drops of ethanol or isopropanol can be added to reduce the surface tension of water-based consolidant. Over the years a number of misting solutions have been proposed and published, both in the conservation literature and at workshops and 'tips and hints' sessions in conferences. There seem to be no off-the-shelf pieces of kit that suit the conservation studio: a home-made tool that can be customised for the object in front of the conservator is common to such reports. Commercial equipment for applying coatings is designed for a well-prepared surface free of contaminants, sometimes aggressively air-blown, or even peened in a form of gentle grit-blasting, at minimum solvent-cleaned to ensure good adhesion between coating and substrate. It is not surprising that many conservators prefer to build their own kit for gentle spray application. There is probably no clear answer to how to do it appropriately, because hand skills and experience with the tool count for so much in these treatments. Once again, surrogate material for practising different strengths of consolidant, types of consolidant, number of applications, possibly with more concentrated solutions of consolidant, can result in successful treatment of objects once regarded as impossible to consolidate. Mimicking the texture, friability, and absorbency of what is to be treated would be much more important here than trying to simulate the age of the object through accelerated ageing.

Some industrial paint spraying methods such as electrostatic spraying have been used to create modern sculptures with a substrate of steel, aluminium, or other metals: the conservation of such pieces after damage to the painted finish on occasion has involved removal of the damaged coating system, preparation of the metal by the same aggressive industrial methods used in the first place, and re-spraying. At times, this is the artist's preferred choice for aesthetic reasons, following discussion with artist, fabricator, owner, and conservator. There are certainly some modern finishes for which traditional consolidation and inpainting of losses are not only challenging but inappropriate for a 'machine-made' aesthetic appearance. There may be no published examples of electrostatic spraying used as a conservation method, but a few such objects might be suitable candidates for an exploration of this option, which could retain what remains of the original electrostatically sprayed coating.

In cases where the object has a *verso*, as for paint on canvas, works on paper, prints, 2D textiles, and the like, a dilute consolidant can be applied from the reverse by spraying, misting, nebulisation, or brushing to provide enhanced security against loss of flaking paint, before local application by brush of the same consolidant to the front is used to secure the larger flakes. Most objects, including wall paintings, present only one surface to the conservator. Access may only be *in situ*, in less than ideal environmental and working conditions, in this case.

The alternative method for consolidating a paint film only accessible from the viewing side is to apply the consolidant in small quantities by brush through a fine fabric or tissue that will not disintegrate when wetted by it, making its removal possible. Nonwoven microfilament fabrics are now used for this purpose, such as **Evolon®**, which is fine, isotropic (meaning that its properties do not vary in different directions), non-stretchy, and resistant to tearing. With some consolidants of near-matching refractive index, it might even be possible to retain small portions of it on the surface being consolidated, encased in the consolidant, for further support.

Temporary consolidation is a method sometimes used on site when the need arises to move an object with a particularly fragile surface indoors for further treatment. The principle of this is to apply (usually by spraying) a material that will vaporise or sublime at (higher) indoor temperatures afterwards, thus removing itself unobtrusively once the move has been completed. The temporary consolidant is acting rather like a mould created to facilitate the excavation and lifting of a fragile object, but without the necessity for manual removal of the mould. For this purpose, **cyclododecane** has been used (Stein *et al.* 2000; Rowe and Rozeik 2008), although there are health concerns over its use: a material intended to vaporise in large quantities from a large object necessarily exposes conservators and other staff to a lot of vapour during its normal use. At the time of writing, few other materials with this unusual range of properties but no potential toxicity have been proposed (Sadek *et al.* 2018).

And what are current research topics in the search for improved consolidants for challenging objects? They certainly include **inorganic and metal oxide coatings** (Centenaro *et al.* 2023) which are a logical development of the use of organosilanes for stone consolidation: the conservation material is designed to have some chemical similarity to the object being treated, to ensure its compatibility and to improve the long-term prospects of the treatment. The other area of development is **functional coatings** (Daehne and Herm 2013) which are often **nanomaterials** (van der Werf *et al.* 2015; Jang and Matero 2018; Bridarolli *et al.* 2020; Chelazzi *et al.* 2024; Gmelch *et al.* 2024; Rawat and Pandey 2024). Nanomaterials generally have distinct properties from their everyday counterparts with the same chemical structure: they have structure on the nano-scale rather than the micro-scale which is 1000 times more coarse, and by careful formulation these properties can be tuned to a desired end-point that can include better solubility characteristics, lower viscosity for better penetration when used for consolidation, better ageing properties, and often a material with water-white clarity. Many classes of nanomaterials being developed for conservation use within large and well-funded projects such as GREENART or GoGreen[2] are aqueous or include water by design, a major driver behind these studies being a reduction in the use of solvents based on petrochemicals. This is a welcome trend that should increase the options available to conservators, for coatings and consolidants and also adhesives.

Notes

1 Even with a very high viscosity like this, bitumen will still flow, though it takes a long time (decades) before it is obvious. See https://en.wikipedia.org/wiki/Pitch_drop_experiment for some illustrations, as noted earlier.

2 See the publications sections for their websites, which in due course will host final reports from these projects that run to 2025 and 2026.

Answers to exercises

Chapter 4

1 The stress is greatest at the smallest cross-section, this is at B.

2 (a) Strain is given by the length of the extension (the length of the light blue, stretched, part of the cylinder) divided by the original length, which is the length of the dark blue cylinder. In this case, all of the rods had the same original length and since that in (a) has clearly stretched the most, it has undergone the greatest strain.

2 (b) Young's modulus is given by stress divided by strain. Since (c) has twice the cross-sectional area of (a), the stress produced in (c) is half that of (a). If the change in length of (c) were half that of (a), the strain in (c) would be half that of (a), and the values of Young's modulus would be the same for both materials. But since the strain in (c) is clearly less than half that of (a) (it is more resistant to deformation), it has the higher Young's modulus.

References

Bridarolli, A., A. Nualart-Torroja, A. Chevalier, M. Odlyha, and L. Bozec. 2020. "Systematic Mechanical Assessment of Consolidants for Canvas Reinforcement Under Controlled Environment." *Heritage Science* 8: 52. https://doi.org/10.1186/s40494-020-00396-x.

Briffa, S. M. and D. A. Vella. 2019. "The Behaviour of As-Applied and Artificially Weathered Silica–Epoxy Consolidants on a Typical Mediterranean Porous Limestone: A Comparison with TEOS." *Heritage Science* 7: 30. https://doi.org/10.1186/s40494-019-0270-1.

Catenazzi, K. 2017. "Evaluation of the Use of *Funori* for Consolidation of Powdering Paint Layers in Wall Paintings." *Studies in Conservation* 62: 96–103. https://doi.org/10.1080/00393630.2015.1131043.

Chelazzi, D., G. Poggi, and P. Baglioni. 2024. "The GREENART Project: Environmentally Friendly and Low Impact Materials for the Conservation of Cultural Heritage." *Studies in Conservation* 69 (sup.1): S25–S34.

Centenaro, S., G. Franceschin, E. Cattaruzza, and A. Traviglia. 2023. "Consolidation and Coating Treatments for Glass in the Cultural Heritage Field: A Review." *Journal of Cultural Heritage* 64: 132–143. https://doi.org/10.1016/j.culher.2023.09.006.

Daehne, A. and C. Herm. 2013. "Calcium Hydroxide Nanosols for the Consolidation of Porous Building Materials: Results from EU-STONECORE." *Heritage Science* 1: 11. http://www.heritagesciencejournal.com/content/1/1/11.

Geiger, T. and F. Michel. 2005. "Studies on the Polysaccharide *JunFunori* Used to Consolidate Matt Paint." *Studies in Conservation* 50: 193–204. https://doi.org/10.1179/sic.2005.50.3.193.

Gmelch, L., E. M. L. D'Emilio, T. Geiger, and C. Effner. 2024. "Degraded Paper: Stabilization and Strengthening Through Nanocellulose Application." *Journal of Paper Conservation*. https://doi.org/10.1080/18680860.2024.2317831.

Graziani, G., E. Sassoni, and E. Franzoni. 2015. "Consolidation of Porous Carbonate Stones by An Innovative Phosphate Treatment: Mechanical Strengthening and Physical-Microstructural Compatibility in Comparison with TEOS-based Treatments." *Heritage Science* 3: 1. https://doi.org/10.1186/s40494-014-0031-0.

Jang, J. and F. G. Matero. 2018. "Performance Evaluation of Commercial Nanolime as a Consolidant for Friable Lime-based Plaster." *Journal of the American Institute for Conservation* 57 (3): 95–111. https://doi.org/10.1080/01971360.2018.1486126.

Kim, E., C. Longo, E. C. Docking, C. Gallupe, R. Langmuir, and C. Magarro. 2024. "Treatment Case Studies in the Consolidation and Adhesion of Cellulose Nitrate." *Studies in Conservation*. https://doi.org/10.1080/00393630.2024.2345422.

Kollaard, N. and W. Wei. 2023. "Adaptation of a Standard Industrial Peel Test for Testing Consolidants in the Conservation Studio." *Journal of the American Institute for Conservation* 62 (2): 139–151. https://doi.org/10.1080/01971360.2022.2130982.

Rawat, A. and S. Pandey. 2024. "Synthesis of Starch Nanoparticles and Its Potential Application as a Green Consolidant for Deteriorated Paper-based Objects." *Studies in Conservation* 69 (sup.1): S298–S304.

Rowe, S. and C. Rozeik. 2008. "The Uses of Cyclododecane in Conservation." *Studies in Conservation* 53 (sup.2): 17–31. https://doi.org/10.1179/sic.2008.53.Supplement-2.17.
Sadek, H., B. H. Berrie, and R. G. Weiss. 2018. "Sublimable Layers for Protection of Painted Pottery during Desalination: A Comparative Study." *Journal of the American Institute for Conservation* 57 (4): 189–202. https://doi.org/10.1080/01971360.2018.1537414.
Stein, R., J. Kimmel, M. Marincola, and F. Klemm. 2000. "Observations on Cyclododecane as a Temporary Consolidant for Stone." *Journal of the American Institute for Conservation* 39 (3): 355–369. https://doi.org/10.1179/019713600806113194.
van der Werf, I. D., N. Ditaranto, R. A. Picca, M. C. Sportelli, and L. Sabbatini. 2015. "Development of a Novel Conservation Treatment of Stone Monuments with Bioactive Nanocomposites." *Heritage Science* 3: 29. https://doi.org/10.1186/s40494-015-0060-3.
van Oosten, T. B. 2022. *Properties of Plastics: A Guide for Conservators*. Los Angeles: Getty Conservation Institute.
Wei, W. 2021. *Art Conservation: Mechanical Properties and Testing of Materials*. Singapore: Jenny Stanford Publishing.

Further reading

Ambers, J. C. H., L. Harrison, and D. Saunders (eds). 2009. *Holding It All Together: Ancient and Modern Approaches to Joining, Repair and Consolidation*. London: Archetype Publications.

Armel, F., L. Dias, J. Mirão, V. Pires, F. Sitzia, S. Martins, M. Costa, and P. Barrulas. 2023. "Chemical Composition of Hydrophobic Coating Solutions and Its Impact on Carbonate Stones Protection and Preservation." *Sustainability* 15 (22): 16135. https://doi.org/10.3390/su152216135.

Artesani, A., F. Di Turo, M. Zucchelli, and A. Traviglia. 2020. "Recent Advances in Protective Coatings for Cultural Heritage – An Overview." *Coatings* 10 (3): 217. https://doi.org/10.3390/coatings10030217.

Barros D'Sa, A., L. Bone, R. Clarricoates, and A. Gent (eds). 2012. *Adhesives and Consolidants in Painting Conservation*. London: Archetype Publications.

Brommelle, N. S., E. M. Pye, P. Smith, and G. Thomson (eds). 1984. *Adhesives and Consolidants*. London: IIC. Available at *Studies in Conservation* 29 (sup.1). https://www.tandfonline.com/loi/ysic20/sic/.

Down, J. L. 2015a. "The Evaluation of Selected Poly(Vinyl Acetate) and Acrylic Adhesives: A Final Research Update." *Studies in Conservation* 60: 33–54. https://doi.org/10.1179/2047058414Y.0000000129.

Down, J. L. 2015b. *Adhesive Compendium for Conservation*. Ottawa: Canadian Conservation Institute.

Feller, R. L. 1994. *Accelerated Aging: Photochemical and Thermal Aspects*. Los Angeles: Getty Conservation Institute.

Ferreira, J. T., A. Bartoletti, S. França de Sá, A. Quye. Y. Shashoua, T. Casimiro, and J. L. Ferreira. 2024. "Proof-of-Concept Study on the Feasibility of Supercritical Carbon Dioxide-Assisted Consolidation Treatment for a Pair of Goalkeeper Gloves on Synthetic Latex-Based Foam Mock-Ups." *Sustainability* 16 (4): 1562. https://doi.org/10.3390/su16041562.

Gherardi, F. D. G., S. Goidanich, A. Colombo, and L. Toniolo. 2017. "On-Site Monitoring of the Performance of Innovative Treatments for Marble Conservation in Architectural Heritage." *Heritage Science* 5: 4. https://doi.org/10.1186/s40494-017-0118-5.

GoGreen Project. 2022–2026. https://www.heritageresearch-hub.eu/project/gogreen/ and https://gogreenconservation.eu/.

Gómez-Sánchez, E., E. Agnini, A. Bridarolli, R. Rolland, S. Kunz, S. Röhrs, C. Harrivelle, E. Casanova, S. Simon, and E. Calparsoro-Forcada. 2022. "Evaluation of the Solubility and Colour Changes of Artificially and Naturally Aged Adhesives for the Conservation of Ceramics and Glass." *Studies in Conservation* 67: 500–517. https://doi.org/10.1080/00393630.2021.1995262.

GREENART Project. 2022–2025. *GREen ENdeavour in Art ResTauration.* https://www.heritageresearch-hub.eu/project/greenart/ and https://www.greenart-project.eu/publications/.

Harrison, K. W. 1979. *Adhesion*, vol. 3. Bristol: Applied Science Publishers.

Horie, C. V. 2010. *Materials for Conservation: Organic Consolidants, Adhesives and Coatings*. Abingdon: Routledge.

Kronthal, L., J. Levinson, C. Dignard, E. Chao, and J. L. Down. 2013. "Beva 371 and Its Use as an Adhesive for Skin and Leather Repairs: Background and a Review of Treatments." *Journal of the American Institute for Conservation* 42 (2): 341–362. https://doi.org/10.1179/019713603806112796.

Kumar, K. and W. S. Ginell. 1997. "A New Technique for Determining the Depth of Penetration of Consolidants into Limestone Using Iodine Vapor." *Journal of the American Institute for Conservation* 36 (2): 143–150. https://doi.org/10.1179/019713697806373181

Li, D. and X. Feigao. 2023. "Conservation of Tortoise Shell Using Hydroxyapatite/Collagen as a Consolidation Material." *Studies in Conservation* 68: 318–325. https://doi.org/10.1080/00393630.2022.2038841.

Messinger. J. M. and P. T. Lansbury. 1989. "Controlling the Refractive Index of Epoxy Adhesives." *Journal of the American Institute for Conservation* 28 (2): 127–136. https://doi.org/10.1179/019713689806046219.

Molina, M. T., E. Cano, and B. Ramírez-Barat. 2023a. "Testing Protective Coatings for Metal Conservation: The Influence of the Application Method." *Heritage Science* 11: 94. https://doi.org/10.1186/s40494-023-00937-0.

Molina, M. T., E. Cano, and B. Ramírez-Barat. 2023b. "Protective Coatings for Metallic Heritage Conservation: A Review." *Journal of Cultural Heritage* 62 (July–August): 99–113.

Molina, M. T., B. Salvadori, E. Cano, D. de la Fuente, and B. Ramírez-Barat. 2023. "Exploration of Coating Alternatives for the Protection of Bare Steel and Brass in Scientific-Technical Artefacts." *Heritage Science* 11: 208. https://doi.org/10.1186/s40494-023-01049-5.

Podany, J., K. M. Garland, W. R. Freeman, and J. Rogers. 2001. "Paraloid B-72 as a Structural Adhesive and as a Barrier within Structural Adhesive Bonds: Evaluations of Strength and Reversibility." *Journal of the American Institute for Conservation* 40 (1): 15–33. https://doi.org/10.1179/019713601806113120.

POPART. *Preservation of Plastic ARTefacts in Museum Collections*. https://popart-highlights.mnhn.fr/.

Ritchie, F. and B. Palumbo. 2023. "Lascaux Adhesives in Objects Conservation: Three Practical Case Studies on Leather, Skin, and Entomological Specimens." *Journal of the American Institute for Conservation* 62 (3): 199–212. https://doi.org./10.1080/01971360.2022.2093538.

Selwitz, C. M. 1988. *Cellulose Nitrate in Conservation*. Los Angeles: Getty Publications.

Smith, C. A., B. J. Lowe, J. Swart, P. Fuentes-Cross, M. Mistral, and R. Te Kanawa. 2022. "Evaluating Five Consolidants for Black-Dyed Māori Textile Artefacts." *Studies in Conservation* 67: 271–288. https://doi.org/10.1080/00393630.2020.1857524.

Soares, I., C. Viana, A. Bartoletti, S. França de Sá, A. Quye, Y. Shashoua, T. Casimiro, and J. L. Ferreira. 2024. "Assessing the Use of Supercritical Carbon Dioxide as a Carrier for Alkoxysilanes to Consolidate Degraded PUR Ester Foams: An Alternative to Traditional Methods." *Sustainability* 16 (11): 4375. https://doi.org/10.3390/su16114375.

Swider, J. R. and M. Smith. 2005. "*Funori*: Overview of a 300-Year-Old Consolidant." *Journal of the American Institute for Conservation* 44 (2): 117–126. https://doi.org/10.1179/019713605806082329.

Tennent, N. H. and J. H. Townsend. 1984. "The Significance of the Refractive Index of Adhesives for Glass Repair." In *Adhesives and Consolidants*, edited by N. S. Brommelle, E. M. Pye, P. Smith, and G. Thomson, 205–211. London: IIC.

Vinçotte, A., E. Beauvoit, N. Boyard, and E. Guilminot. 2019. "Effect of Solvent on Paraloid® B-72 and B-44 Acrylic Resins used as Adhesives in Conservation." *Heritage Science* 7: 42. https://doi.org/10.1186/s40494-019-0283-9.

Walsh, Z., E. Janeček, M. Jones, and O. A. Scherman. 2017. "Natural Polymers as Alternative Consolidants for the Preservation of Waterlogged Archaeological Wood." *Studies in Conservation* 62: 173–183. https://doi.org/10.1179/2047058414Y.0000000149.

Zhao, J., H. Luo, L. Wang, W. Li, T. Zhou, and B. Rong. 2013. "TEOS/PDMS-OH Hybrid Material for the Consolidation of Damaged Pottery." *Heritage Science* 1: 12. http://www.heritagesciencejournal.com/content/1/1/12.

Index